AF658328

Fermented fruits and vegetables

A global perspective

by
Mike Battcock
and
Sue Azam-Ali
Intermediate Technology
Schumacher Centre for Technology and Development
Bourton on Dunsmore, Rugby, Warwickshire, UK

Reprinted by arrangement with the
Food and Agriculture Organization of the United Nations
by
Daya Publishing House®
A Division of
Astral International Pvt. Ltd.
New Delhi - 110 002

"This book was originally published by the Food and Agriculture Organization of the United Nations (FAO) as Fermented fruits and vegetables-A global perspective, FAO Agricultural services bulletin, 134".

"The designations employed and the presentation of material in this publication do not imply the expression of any opinion, whatsoever on the part of the Food and Agriculture Organization of the United Nations concerning the legal status of any country, territory, city or area or of its authorities, or concerning the delimitations of its frontiers or boundaries."

'The designations 'developed' and 'developing' economies are intended for statistical convenience and do not necessarily express a judgment about the stage reached by a particular country, territory or area in the development process.'

© Copyright 1998 by FAO
First Indian Edition, 2001
Reprinted, 2016
ISBN: 978-81-7035-248-8

Published by : **Daya Publishing House®**
A Division of
Astral International Pvt. Ltd.
– ISO 9001:2008 Certified Company –
4760-61/23, Ansari Road, Darya Ganj
New Delhi-110 002
Ph. 011-43549197, 23278134
E-mail: info@astralint.com
Website: www.astralint.com

Printed at : **Chawla Offset Printers,**
Delhi-110 052

FOREWORD

Fermentation is an affordable and manageable food preservation technique, which is very appropriate in use where other food processing technologies such as canning and freezing are either inaccessible or unavailable. The fermentation of staples serves as a major source of nourishment for large populations in rural communities and contributes significantly to food security by increasing the range of raw materials, which can be used in the production of edible products.

The consolidated information base on traditional small-scale food fermentations is however very weak and existing information on the subject is widely dispersed. Furthermore, "indigenous knowledge"on fruit and vegetable fermentations is being lost as technologies evolve and populations move away from traditional food preservation practices. The Agro-Industries and Post-Harvest Management Service (AGSI) of FAO has therefore initiated the publication of a document series on fermented foods. The present publication on fermented fruits and vegetables is the first of the series.

It is hoped that this and other documents in the series will generate wider interest in developing and improving traditional small-scale food fermentation technologies.

M. Satin
Chief,
Agro-Industries and Post-Harvest Management Service

EXECUTIVE SUMMMARY

1. Fermentation is one of the oldest forms of food preservation technology in the world.

2. Fermented foods are popular throughout the world.

3. Fermented foods make a significant contribution to the diet of millions of individuals.

4. Fermentation is a relatively efficient, low energy preservation process, which increases the shelf life and decreases the need for refrigeration or other form of food preservation technology. It is therefore a highly appropriate technique for use in developing countries and remote areas.

5. There is tremendous scope and potential for the use of micro-organisms towards meeting the growing world demand for food, through efficient utilisation of available natural food and feed stocks and the transformation of waste materials.

6 There is a danger that the introduction of 'western foods' with their glamorous image will displace these traditional fermented foods.

7. It is essential that the knowledge of fermented food production is not lost.

8. It is essential to increase the knowledge and understanding of the methods of preparation, in order to improve the efficiency of fermentation, especially the traditional processes as the yields of traditional fermentation processes are often low and sometimes the products are unsafe.

9. For fermented products such as cheese, bread, beer and wine, which are produced on a commercial scale, a good understanding of the microbial processes has been developed. However with many of the fermented products in Africa, Asia and Latin America, knowledge of the process is poor. It is likely that the basic principles apply across the board, but production conditions vary enormously from region to region, giving rise to numerous variations of the basic fermented product. It is not the intention or the desire to standardise the process and thereby lose this huge diversity, rather it is to harness the tremendous potential these methods have to contribute to increasing not only the quantity, but quality of food available to the worlds population.

10. The collection and preservation of indigenous knowledge is of interest to governments, historians, anthropologists and scientists, to name but a few. Several individuals and organisations are actively involved in research in this area.

11. Most traditional fermented products are made by natural fermentations carried out in a non-sterile environment. The specific environmental conditions cause a gradual selection of microorganisms responsible for the desired final product. This is appropriate for small-scale production for home consumption. However the method is difficult to control and there are risks of accompanying micro-flora causing spoilage and unsafe products.

12. The art of traditional processes needs to be refined to incorporate objective methods of process control and to standardise quality of the final product without losing their desirable attributes such as improved keeping quality, taste and nutritional qualities.

13. Once the details of the fermentation process and the microbes involved are known and understood, it is possible to begin to refine and improve the process.

14. The areas where the efficiency and yield of food fermentation processes can be increased are: the selection or development of more productive microbial strains; the control and manipulation of culture conditions and the improvement of product purification and concentration.

15. It is often felt that traditional products made at the small scale are unhygienic and unsafe. This is sometimes true. However the case is often overstated. Many fermented foods are inherently safe due to low moisture contents or high acidity.

16. Quality control procedures are essential for the production of safe products and contribute to the success of small food processing businesses. Appropriate quality control procedures need to be developed and implemented. These procedures need to be developed with the processors who must understand and apply them.

17. Documentation of the traditional methods of food fermentation and research to identify improved methods of production are meaningless if the results are not disseminated to those who are likely to put them into practice. There is a danger of mystifying the fermentation process by enrobing it in scientific theory.

18. Fermented foods often are considered as poor man's food. As soon as a family can afford to, they move away from carrying out home fermentation. This is a pity, because as we have seen earlier, fermented food products have many nutritional advantages, which surpass western-style fast foods and processed foods.

19. Fermented foods should be recognised as part of each countries heritage and culture and efforts made to preserve the methods of production. A recognised body (government or non-government) should take the responsibility for the collection of details and the promotion of fermented food products.

20. Consumers need to be made aware of the numerous benefits of fermented foods and their prejudices against fermented foods, especially those traditionally produced at the home scale, dispelled

AUTHORS

Mike Battcock

Senior Agro-processing Specialist,
Intermediate Technology, Schumacher Centre for Technology and Development, Bourton Hall, Bourton on Dunsmore, Rugby, Warwickshire, CV23 9QZ, UK

Dr Sue Azam-Ali,

Agro-processing Specialist,
Intermediate Technology, Schumacher Centre for Technology and Development, Bourton Hall, Bourton on Dunsmore, Rugby, Warwickshire, CV23 9QZ, UK

ACKNOWLEDGEMENTS

Cover Photograph: "Preparation of Mangoes for Pickling" by Sue Azam-Ali, Intermediate Technology.

TABLE OF CONTENTS

Page

LIST OF TABLES

INTRODUCTION

Agricultural crops are processed for many different reasons. These range from the removal of anti-nutritional components and increasing the storage life of the final product to adding value to increase both employment and income generating opportunities. Fermentation is one of the most ancient and most important food processing technologies. However scientists and policy makers have neglected this area, particularly traditional fermented products from developing countries.

Fermented foods: an ancient tradition

Fermentation is one of the oldest forms of food preservation technologies in the world. Indigenous fermented foods such as bread, cheese and wine, have been prepared and consumed for thousands of years and are strongly linked to culture and tradition, especially in rural households and village communities.

The development of fermentation technologies is lost in the mists of history. Anthropologists have suggested that it was the production of alcohol that motivated primitive people to settle down and become agriculturists. Some even think the consumption of fermented food is pre-human (Stanton, 1985)[1]. The first fermented foods consumed probably were fermented fruits. Hunter-gatherers would have consumed fresh fruits but at times of scarcity would have eaten rotten and fermented fruits. Repeated consumption would have led to the development of the taste for fermented fruits. There is reliable information that fermented drinks were being produced over 7,000 years ago in Babylon (now Iraq), 5,000 years ago in Egypt, 4,000 years ago in Mexico and 3,500 years ago in Sudan (Dirar, 1993)[2], (Pedersen, 1979)[3].

Bread-making probably originated in Egypt over 3,500 years ago (Sugihara, 1985)[4]. Several triangular loaves of bread have been found in ancient tombs. Fermentation of milk started in many places with evidence of fermented products in use in Babylon over 5,000 years ago. There is also evidence of fermented meat products being produced for King Nebuchadnezer of Babylon. China is thought to be the birthplace of fermented vegetables and the use of *Aspergillus* and *Rhizopus* moulds to make food. The book called "Shu-Ching" written in the Chou dynasty in China (1121-256 BC) refers to the use of "chu" a fermented grain product (Yokotsuka, 1985)[5].

Knowledge about traditional fermentation technologies has been handed down from parent to child, for centuries. These fermented products have been adapted over generations; some products and practices no doubt fell by the wayside. Those that remain today have not only survived the test of time but also more importantly are appropriate to the technical, social and economic conditions of the region.

Fermented foods are culturally and economically important

Fermentation is a relatively efficient, low energy preservation process, which increases the shelf life and decreases the need for refrigeration or other forms of food preservation technology. It is therefore a highly appropriate technique for use in developing countries and remote areas where access to sophisticated equipment is limited. Fermented foods are popular throughout the world and in some regions make a significant contribution to the diet of millions of individuals.

In Asia the preparation of fermented foods is a widespread tradition. The fermented products supply protein, minerals and other nutrients that add variety and nutritional fortification to otherwise starchy, bland diets. For instance Soy sauce is consumed throughout the world and is a fundamental ingredient in diets from Indonesia to Japan. Over one billion litres are produced each year in Japan alone. "*Gundruk*" which is a fermented and dried vegetable product is very important for ensuring food security for many Nepali communities especially in remote areas. It is served as a side dish with the main meal and is also used as an appetiser in the bland, starchy diet. The annual production of *gundruk* in Nepal is estimated at 2,000 tons. *Gundruk* is an important source of minerals particularly during the off-season when the diet consists primarily of starchy tubers and maize, which tend to be low in minerals. In Africa fermented cassava products (like *gari* and *fufu*) are a major component of the diet of more than 800 million people and in some areas these products constitute over 50% of the diet.

The need for research

Although fermentation of foods has been in use for thousands of year, it is likely that the microbial and enzymatic processes responsible for the transformations were largely unknown. It is only recently that there has been a development in the understanding of these processes and their adaptation for commercialisation. There is tremendous scope and potential for the use of micro-organisms towards meeting the growing world demand for food, through efficient utilisation of available natural food and feed stocks and the transformation of waste materials.

Because of the tremendously important role indigenous fermented fruits and vegetables play in food preservation and their potential to contribute to the growing food needs of the world, it is essential that the knowledge of their production is not lost. There is a danger that the introduction of 'western foods' with their glamorous image will displace these traditional foods.

This book presents an overview of the fermented fruit and vegetable products of Africa, Asia and Latin America. The bulletin aims to introduce the reader to the vast wealth of knowledge, much of which is indigenous and undocumented and the importance attached to fermented fruits and vegetables in the diet. At the same time it is a practical handbook, allowing those who are interested to reproduce the products.

DESCRIPTION OF TERMS USED

Fermentation

Fermentation is the "slow decomposition process of organic substances induced by micro-organisms, or by complex nitrogenous substances (enzymes) of plant or animal origin" (Walker, 1988)[6]. It can be described as a biochemical change, which is brought about by the anaerobic or partially anaerobic oxidation of **carbohydrates** by either microorganisms or enzymes. This is distinct from putrefaction, which is the degradation of **protein** materials.

The changes caused by fermentation can be both advantageous and disadvantageous. Fermentation, initiated by the action of microorganisms occurs naturally and is often part of the process of decay, especially in fruits and vegetables. However, fermentation can be controlled to give beneficial results. Fermentation is a relatively efficient, low energy preservation process, which increases the shelf life and decreases the need for refrigeration or other form of food preservation technology. It is therefore a highly appropriate technique for use in developing countries and remote areas where access to sophisticated equipment is limited.

Fruits

There are several definitions of "fruit", which makes classification and distinction between fruit and vegetable difficult. The everyday usage of the word "fruit" defines fruit as "The edible product of a plant or tree, consisting of the seed and its envelope, especially the latter when juicy and pulpy" (Little *et al*, 1973)[7]. The scientific definition of a fruit is "The structure that develops from the ovary of an angiosperm as the seeds mature, with (false fruit) or without (true fruit) associated structures" (Walker, 1988)[8]

In terms of food processing, fruits are nearly all acidic and are therefore called 'high acid' foods. The acidity naturally controls the type of organism that can grow in fruits, with yeasts and moulds being the only spoilage organisms likely to be found on fruit products. The acidity level of tropical fruits, such as banana, mango and papaya, decreases as the fruit ripens (Anon, 1993)[9]. With respect to food processing and preservation, it is probably this definition of a fruit that will be most useful.

Vegetables

A vegetable is "a plant cultivated for food, especially an edible herb or root used for human consumption"(Little *et al,* 1973)[10]. In general, vegetables tend to be less sweet than fruits and often require some form of processing to increase their edibility.

In terms of food processing, vegetables are classified as 'low acid' foods due to their lower levels of acidity. Low acid foods are more prone to deterioration by microorganisms and can in fact provide an ideal substrate for food poisoning organisms when in a moist environment. Low acid foods can be safely preserved by making them more acidic, either through pickling or salting or drying (Anon, 1993)[11].

Agro-processing

Agro-processing describes the transformation of agricultural produce into a different physical or chemical state. The term *agro* encompasses a wide range of food and non-food agricultural products. The term *food processing* applies only to products, which are suitable for human consumption. *Agro processing* applies to any of the numerous activities that take place in the chain of events between harvest or slaughter of the raw material and production of the final product. It covers a range of processes with varying degrees of complexity and technical input to suit the individual situation. Different treatments range from the relatively simple processes of rice husking, drying and grinding to the more complex transformation of oilseeds into margarine and essential oil distillation.

Salometer

Salt present in a brine is expressed as degrees "Salometer" which is a percent saturation of sodium chloride by weight. A saturated solution of pure sodium chloride (100 degrees Salometer) contains 26.359g per 100ml at 15.5 degrees C. Therefore a 10 degree Salometer contains 2.64% salt by weight.

Biotechnology

Biotechnology can be described as the application of scientific and engineering principles to the processing of materials, for the provision of goods and services, through the use of biological systems and agents (Anon, 1992)[12].

Complex carbohydrate

A complex carbohydrate is one that is composed of long branched chains of single sugar units (glucose, fructose or galactose). Examples of complex carbohydrates include starch and cellulose.

Reducing sugar

A reducing sugar is a sugar, which has reactive aldehyde or ketone groups. All simple sugars are reducing sugars. Sucrose, a common sugar, is not a reducing sugar.

Micro-organism/Microbe

Microbe and microorganism are generic terms for the group of living organisms, which are microscopic in size. Included in the definition are bacteria, viruses, moulds, yeasts and fungi.

Enzyme

An enzyme is a biological catalyst, which is used to facilitate and speed up reactions. It is a protein and requires a specific substrate to work on. Its working conditions are set within narrow limits – for example, optimum temperature, pH conditions and oxygen concentration. At temperatures above 42°C, mammalian proteins (and therefore enzymes) are denatured. However, certain bacterial enzymes are tolerant of a more diverse temperature range.

Hydrolysis

Hydrolysis is the splitting or breaking down of complex molecules by the action of enzymes or acid. For example the hydrolysis of starch and cellulose both yield simple glucose units.

Aerobic

With reference to microorganisms, an aerobic organism is one, which requires oxygen for survival.

Anaerobic

An anaerobic organism is one which does not require oxygen for survival.

Facultative anaerobe

A microorganism which can adapt to exist with or without oxygen.

Microaerophilic

An organism which requires small amounts of oxygen for survival.

Water activity (a_w)

Water exists in two states within a cell – free and bound. Water activity is a measure of the amount of free water available for a potential reaction – microbial or enzymatic. Water activity is a measure of the free moisture in a product and is the quotient of the water vapour pressure of the substance divided by the vapour pressure of pure water at the same temperature. It is measured on a scale of 0 to 1.0 where 1.0 is the activity of pure water.

pH

pH is a measure of the hydrogen ion concentration. It is measured on a scale of 1 to 14, where 1 represents a high concentration of hydrogen ions (acidic) and 14 represents a low concentration (alkaline). The optimum pH for most microorganisms is near neutral (pH 7.0). However, certain species are acid tolerant . Foods with a pH of 4.6 or lower are termed high acid foods. If the pH is above 4.6, it is a low acid food and is more prone to bacterial and fungal spoilage.

CHAPTER 1

THE BENEFITS OF FERMENTING FRUITS AND VEGETABLES

Fermenting fruits and vegetables can bring many benefits to people in developing countries. Fermented foods play an important role in providing food security, enhancing livelihoods and improving the nutrition and social well being of millions of people around the world, particularly the marginalised and vulnerable.

1.1 Improving food security

Eight hundred million people do not have enough food to eat. If we include those not free from hunger the figure rises to 1.2 billion people. This is one fifth of the World's population. A further two billion people are deficient in one or more micronutrients (Anon, 1996)[13]. In the seventies, food security was viewed mainly in terms of food supply at the global and national levels. Since then there has been a major shift in understanding of food security with more emphasis on access to food rather than purely on production. The Food and Agriculture Organisation of the United Nations (FAO), amongst other influential organisations, has recognised that the problem of food security cannot be tackled in isolation. Moreover that it is an integral component of other development issues. FAO highlights the fact that the world food insecurity problem is a result of undemocratic and inequitable distribution of and access to resources rather than a problem of global food production (Anon, 1995)[14], (Anon, 1996)[15].

Fermentation technologies play an important role in ensuring the food security of millions of people around the world, particularly marginalised and vulnerable groups. This is achieved through improved food preservation, increasing the range of raw materials that can be used to produce edible food products and removing anti- nutritional factors to make food safe to eat.

1.1.1 Food preservation

Fermentation is a cheap and energy efficient means of preserving perishable raw materials. When harvested, fruit and vegetables, undergo rapid deterioration, especially in the humid tropics where the prevailing environmental conditions accelerate the process of decomposition. There are several options for preserving fresh fruit and vegetables including drying, freezing, canning and pickling. However many of these are inappropriate for use on the small-scale in developing countries. For instance the canning of vegetables at the small-scale has serious food safety implications and contamination with botulism is a possibility. Freezing of fruits and vegetables is not economically viable at the small-scale. Fermentation requires very little sophisticated equipment, either to carry out the fermentation or for subsequent storage of the fermented product. It is a technique that has been employed for generations to preserve food for consumption at a later date and to improve food security. There are examples from around the world of the role fermented foods have played in preserving food to enhance food security.

Fermented foods for survival in Sudan

About 60% of the fermented foods of Sudan are famine or survival foods. Many of the fermented foods have been developed in Western Sudan in the Kordofan and Darfur regions, which are traditional famine areas. The strong link between fermented foods and food shortages is revealed by the fact that when a family becomes rich a number of fermented foods are no longer prepared. The techniques used are very effective methods of food preservation. The products can be preserved for years through the double action of fermentation itself (which produces anti-microbial acids) and sun drying. Sudan is probably the hottest and driest country in Africa. Through the years women have made full use of this free solar energy. Shade temperatures in the summer reach 45-50°C and the hot sands outside the shade reach more than 70°C. Dried and fermented foods together with the seeds and fruits that can be gathered from the wild have saved lives especially those of children in the past and in the present in times of shortage (Dirar, 1992)[16]. During the 1983-85 famine, relief workers found that people had survived by producing specific traditional fermented food products, especially *Kawal* (Arthur, 1986)[17].

***Gundruk*: an important fermented product in Nepal**

Gundruk is a fermented and dried vegetable product. It is produced by shredding the leaves of mustard, radish and cauliflower leaves and placing them in an earthenware pot to ferment. After five to seven days the leaves are removed and dried in the sun. Gundruk is a very important food product in Nepal ensuring food security for many Nepali communities especially in remote areas. It is served as a side dish with the main meal and is also used as an appetiser in the bland, starchy diet. The annual production of *gundruk* in Nepal is estimated at 2,000 tons and most of the production is carried out at the household level. *Gundruk* is also an important source of minerals particularly during the off-season when the diet consists of mostly starchy tubers and maize which tend to be low in minerals (Karki, 1986)[18].

1.1.2 Salvaging waste foods

Fermentation can salvage waste food which otherwise would not be usable as food by changing the consistency of the product and making it digestible. This increases the range of raw materials available as food.

Bones and hides

A wide range of "waste" products is fermented to produce edible food products in Sudan. These include bones, hides and locusts. Fresh bones are fermented into a variety of products. "*Dodery*" is produced by chopping bones into small pieces and placing them into fermenting vats. They are subsequently covered in water, left for three days, removed, crushed into a paste and mixed with the ash from burnt sorghum stalks. The mixture is returned to the fermenting vat for a further two to five days. The final product is rolled into balls and has a shelf like of up to two months. Another product "*Kaidu digla*" is made from the vertebrae of the backbone. These are chopped ınto smaller pieces and then sun-dried. After drying they are pounded with stones; mixed with water and salt; moulded into balls and allowed to ferment (Dirar, 1992)[19].

Use of waste products in Indonesia

In Indonesia a variety of waste products are fermented to produce nutritious food products. *Tempe-bongrek* is a protein rich food made in Indonesia by fermenting peanut and coconut press-cake, remaining after oil extraction. The product is similar to traditional tempeh produced from the fermentation of soya beans. The production of *tempeh- bongrek* is a mould fermentation, initiated by inoculation of the soaked, acidified press-cake with *Rhizopus* species. The inoculated cakes are placed on banana leaves or plastic sheets in a dark room for about 2 days. An incubation temperature of 37°C is optimal for the mould and prevents the growth of *P. cocovenenans*, which produces *bongkrek* toxin. A pH of less than 6.0 also prevents the development of *bongkrek* toxin. *Ontjom* is produced from waste groundnut press cake, tapioca waste and the solid waste of *tahu. Ontjom* is prepared using a mixed culture of microorganisms with *Rhizopus* or *Neurospora* species predominating. *Ontjom* is mainly produced in west Java where it is consumed as a side dish in the form of deep fried slices. It forms an important daily food item for the west Javanese, particularly those from the lower income groups. Fresh coconut residue, left over from the production of coconut cream or milk, can be fermented by *Bacillus subtilis*, in an alkaline fermentation, to produce *semayi*, which is widely consumed in Indonesia (Steinkraus, 1996)[20].

Pineapple peel vinegar

A considerable amount of the fruit can be wasted in the peeling and preparation of pineapples in Latin America. Through fermentation, a product can be produced from the peelings, that would otherwise have been discarded. The peelings are placed in containers of water and sugar and yeast are added. They are allowed to sit for about 8 days, after which pineapple peel vinegar is produced. The product is of a distinct, light pineapple flavour and can be used in the same way as other vinegars.

1.1.3 Removal of anti-nutritional factors

Many fruits and vegetables contain naturally occurring toxins and anti-nutritional compounds. These can be removed or detoxified by the action of microorganisms during fermentation. For instance the fermentation process that produces the Sudanese product *Kawal* removes the toxins from the leaves of *Cassia obtusifolia* and fermentation is an important step in ensuring that cassava is safe to eat.

Removing cyanide by fermentation

Cassava contains a naturally occurring chemical: cyanogenic glucoside. When eaten raw or improperly processed, this substance releases cyanide into the body, which can be fatal. Correct processing removes this chemical. The cassava is first peeled (as about 60-70% of the poison is in the peel) and then soaked in stagnant water or fermented in sacks for about three days. It is sometimes grated or rasped as this helps to speed up the fermentation process. At the beginning of the fermentation, *Geotricum candida* acts on the cassava. This tends to make the product acidic, which finally kills off the micro-organisms as they cannot exist in such a medium. A second strain of microorganisms *(Cornibacterium lactii)*, which can tolerate the acidic environment then take over and by the third day 90-95% of the dangerous chemical will have been hydrolysed. The cassava also develops its characteristic flavour. The product is then sieved and the fine starch particles are fried in an iron pan alone over a flame or with some palm oil. During this process most, if not all the remaining toxins are given off. The liquor from a previous fermentation is used as a starter, thereby reducing the period of fermentation to about 6-8 hours.

1.2 Increasing income and employment

The production of fermented fruit and vegetable products provides income and employment to millions of people around the world.

Food processing is probably the most important source of income and employment in Africa, Asia and Latin America. The Food and Agriculture Organisation of the United Nations has stated that value added through marketing and processing raw products can be much greater than the value of primary production (Anon, 1995)[21]. For instance in sub-Saharan Africa more than 60% of the workforce is employed in the small scale food processing sector, and between one third and two thirds of value added manufacturing is based on agricultural raw materials (World Bank, 1989)[22], (Conroy et al, 1995)[23]. This is particularly important, as agriculture and the formal sector are unable to absorb the growing labour force in many countries.

Fermented foods are popular throughout the world and the production of fermented food products is important in many countries in providing income and employment.

In Africa, fermented cassava products (like *Gari* and *Fufu)* are a major component of the diet of more than 800 million people and in some parts of Africa it constitutes over 50% of the diet (Oyewole, 1992)[24]. In Asia the preparation of fermented foods is a widespread tradition. *Kimchi* (a fermented cabbage product) is the major food product of Korea. Soy sauce (a fermented legume product) is economically important from Indonesia to Japan. Over a billion litres are produced each year in Japan alone. Over 2000 million litres are produced each year in Korea and over 150 million litres in Taiwan. *Miso* (a fermented legume product) is also very important in Asia with over 560,000 tons produced a year in Japan alone (Anon, 1982)[25]. In Latin America, fermented cereal products, alcoholic drinks and fermented milk products are three of the most important sectors of the economy.

1.3 <u>Improving nutrition</u>

The optimum health and nutrition of individuals is dependent upon a regular supply of food and a balanced diet. When diets are sub-optimal, the individual's capacity for work and achievements are greatly reduced. The most vulnerable groups are women, children and weaning infants. Availability of food, dietary restrictions and taboos, misconceptions, limited time available for feeding or eating compound to create a group of individuals who are nutritionally disadvantaged. Approximately 30% of women consume less than their daily requirements of energy and at least 40% of women worldwide suffer from iron-deficiency anaemia. Fermentation can enhance the nutritional value of a food product though increased vitamin levels and improved digestibility.

1.3.1 Vitamins

Fermentation processes can result in increased levels of vitamins in the final product. *Saccharomyces cerevisiae* is able to concentrate large quantities of thiamine, nicotinic acid and biotin and thus form enriched products.

- Sorghum beer in Southern Africa contains relatively high levels of riboflavin and nicotinic acid, which are important for people consuming a high maize diet. Pellagra (a vitamin deficiency disease associated with high maize diets) is unusual in communities in which sorghum beer is consumed. Even children benefit from consuming the dregs, which contain relatively little alcohol but are rich in vitamins.
- Palm wine in West Africa is high in vitamin B12, which is very important for people with low meat intake, and who subsist primarily on a vegetarian diet.
- *Pulque* (a fermented plant sap) is an important source of vitamins for the economically deprived in Mexico. The fermentation process involved in *Pulque* production increases its vitamin content. For instance the vitamin content (milligrams of vitamins per 100g of product) of pulque increases from 5 to 29 for thiamine, 54 to 515 for niacin and 18 to 33 for riboflavin (Steinkraus, 1992) [26] during fermentation.
- *Idli* (a lactic acid bacteria fermented product consumed in India) is high in thiamine and riboflavin.

1.3.2 Digestibility

Microorganisms contain certain enzymes, such as cellulases, which are incapable of being synthesised by humans. Microbial cellulases hydrolyse cellulose into sugars which are then readily digestible by humans. Similarly pectinases soften the texture of foods and liberates sugars for digestion. Fermented foods are often more easily digestible than unfermented foods (Kovac, 1997)[27], (Parades-Lopez, 1992)[28].

Lactic acid fermented weaning foods are traditionally produced in developing countries, both to improve the safety of the food and to improve its digestibility. Starchy porridges are commonly fed to weaning infants in developing countries. The consistency of these gruels, combined with the small capacity of the infants stomach, means that it is physically impossible for the child to consume adequate energy to meet its high demands. By acidifying the porridge through lactic acid fermentation, starch is hydrolysed into shorter chains of glucose and dextrose, which reduce the viscosity of the porridge and increase its energy density. Thus the child is more able to meet its energy requirements.

1.4 Medicinal benefits

There are many traditional beliefs about the medicinal properties of fermented food products. The Fur ethnic group in Sudan strongly believe that the consumption of fermented foods protects them from disease (Dirar, 1992)[29]. *Koumiss* (a fermented milk product in Russia) has been used to treat tuberculosis. *Pulque* (a fermented fruit sap) is felt to have medicinal properties in Mexico.

There is a sound scientific basis to these assertions:

- The lowering of the pH inhibits the growth of food spoiling or poisoning bacteria and destroys certain pathogens (Hammes, and Tichaczek, 1994)[30].
- Certain lactic acid bacteria (e.g. *Lactobacillus acidophilus*) and moulds have been found to produce antibiotics and bacteriocins (Wood and Hodge, 1985)[31] (Matususaki *et al*, 1997)[32] (Adams and Nicolaides, 1997)[33], (Gourama and Bullerman, 1995)[34], (Nout, 1995)[35].
- The beneficial health effects of lactic acid bacteria on the intestinal flora are well-documented (Ottogalli and Galli, 1997)[36], (Motarjemi *et al*, 1996)[37].
- Substances in fermented foods have been found to have a protective effect against the development of cancer (Frohlich *et al*, 1997)[38].

Fermentation is a traditional method of reducing the microbial contamination of porridges in Kenya (Watson, Ngesa, Onyang, Alnwick and Tomkins, 1996)[39]. A study in Tanzania has shown that children fed with fermented gruels had a 33% lower incidence of diarrhoea than those fed unfermented gruels, owing to the inhibition of pathogenic bacteria by lactic acid forming bacteria (Svanberg, 1992)[40].

1.5 Improving cultural and social well being

Fermentation can improve the flavour and appearance of food. One important area is the creation of meat-like flavour. Over the years, Sudanese women have developed products to replace meat in their diets. These include "*kawal*", fermented wild legume leaves, "*sigda*" (fermented sesame press-cake) and "*furundu*" (fermented red sorrel seeds). The strong flavours of fermented food products can enhance a dull diet. Fermented vegetables such as pickles, *gundruk* and sauerkraut are used as condiments to enhance the overall flavour of the meal. A small amount of pickle can make a bland starchy diet (like dahl and rice in Asia) much more appealing (Battcock, 1992)[41].

CHAPTER 2

BASIC PRINCIPLES OF FERMENTATION

2.1 The diversity of fermented foods

Numerous fermented foods are consumed around the world. Each nation has its own types of fermented food, representing the staple diet and the raw ingredients available in that particular place. Although the products are well know to the individual, they may not be associated with fermentation. Indeed, it is likely that the methods of producing many of the worlds fermented foods are unknown and came about by chance. Some of the more obvious fermented fruit and vegetable products are the alcoholic beverages - beers and wines. However, several more fermented fruit and vegetable products arise from lactic acid fermentation and are extremely important in meeting the nutritional requirements of a large proportion of the world's population. Table 2.1 contains examples of fermented fruit and vegetable products from around the world.

2.2 Organisms responsible for food fermentations

The most common groups of microorganisms involved in food fermentations are:

- Bacteria
- Yeasts
- Moulds

2.2.1 Bacteria

Several bacterial families are present in foods, the majority of which are concerned with food spoilage. As a result, the important role of bacteria in the fermentation of foods is often overlooked. The most important bacteria in desirable food fermentations are the *lactobacillaceae* which have the ability to produce lactic acid from carbohydrates. Other important bacteria, especially in the fermentation of fruits and vegetables, are the acetic acid producing *acetobacter* species.

2.2.2 Yeasts

Yeasts and yeast-like fungi are widely distributed in nature. They are present in orchards and vineyards, in the air, the soil and in the intestinal tract of animals. Like bacteria and moulds, yeasts can have beneficial and non-beneficial effects in foods. The most beneficial yeasts in terms of desirable food fermentation are from the *Saccharomyces* family, especially *S. cerevisiae*. Yeasts are unicellular organisms that reproduce asexually by budding. In general, yeasts are larger than most bacteria. Yeasts play an important role in the food industry as they produce enzymes that favour desirable chemical reactions such as the leavening of bread and the production of alcohol and invert sugar.

Table 2.1 Fermented foods from around the world.

Name and region	Type of product
Indian sub-continent	
Acar, Achar, Tandal achar, Garam nimboo achar	Pickled fruit and vegetables
Gundruk	Fermented dried vegetable
Lemon pickle, Lime pickle, Mango pickle	
South East Asia	
Asinan, Burong mangga, Dalok, Jeruk, Kiam-chai, Kiam-cheyi, Kong-chai, Naw-mai-dong, Pak-siam-dong, Paw-tsay, Phak-dong, Phonlami-dong, Sajur asin, Sambal tempo-jak, Santol, Si-sek-chai, Sunki, Tang-chai, Tempoyak, Vanilla,	Pickled fruit and vegetables
Bai-ming, Leppet-so, Miang	Fermented tea leaves
Nata de coco, Nata de pina	Fermented fruit juice
East Asia	
Bossam-kimchi, Chonggak-kimchi, Dan moogi, Dongchimi, Kachdoo kigactuki, Kakduggi, Kimchi, Mootsanji, Muchung-kimchi, Oigee, Oiji, Oiso baegi, Tongbaechu-kimchi, Tongkimchi, Totkal kimchi,	Fermented in brine
Cha-ts'ai, Hiroshimana, Jangagee, Nara senkei, Narazuke, Nozawana, Nukamiso-zuke, Omizuke, Pow tsai, Red in snow, Seokbakji, Shiozuke, Szechwan cabbage, Tai-tan tsoi, Takana, Takuan, Tsa Tzai, Tsu, Umeboshi, Wasabi-zuke, Yen tsai	Pickled fruit and vegetables
Hot pepper sauce	
Africa	
Fruit vinegar	Vinegar
Hot pepper sauce	
Lamoun makbouss, Mauoloh, Msir, Mslalla, Olive	Pickled fruit and vegetables
Oilseeds, Ogili, Ogiri, Hibiscus seed	Fermented fruit and vegetable seeds
Wines	Fermented fruits
Americas	
Cucumber pickles, Dill pickles, Olives, Sauerkraut,	Pickled fruit and vegetables
Lupin seed, Oilseeds,	Pickled oilseed
Vanilla, Wines	Fermented fruit and vegetable
Middle East	
Kushuk	Fermented fruit and vegetables
Lamoun makbouss, Mekhalel, Olives, Torshi, Tursu	Pickled fruit and vegetables
Wines	Fermented fruits
Europe and World	
Mushrooms, Yeast	Moulds
Olives, Sauerkohl, Sauerruben	Pickled fruit and vegetables
Grape vinegar, Wine vinegar	Vinegar
Wines, Citron	Fermented fruits

(Taken from G Campbell-Platt (1987))[42]

2.2.3 Moulds

Moulds are also important organisms in the food industry, both as spoilers and preservers of foods. Certain moulds produce undesirable toxins and contribute to the spoilage of foods. The *Aspergillus* species are often responsible for undesirable changes in foods. These moulds are frequently found in foods and can tolerate high concentrations of salt and sugar. However, others impart characteristic flavours to foods and others produce enzymes, such as amylase for bread making. Moulds from the genus *Penicillium* are associated with the ripening and flavour of cheeses. Moulds are aerobic and therefore require oxygen for growth. They also have the greatest array of enzymes, and can colonise and grow on most types of food. Moulds do not play a significant role in the desirable fermentation of fruit and vegetable products.

When microorganisms metabolise and grow they release by-products. In food fermentations the by-products play a beneficial role in preserving and changing the texture and flavour of the food substrate. For example, acetic acid is the by-product of the fermentations of some fruits. This acid not only affects the flavour of the final product, but also more importantly has a preservative effect on the food. For food fermentations, microorganisms are often classified according to these by-products. The fermentation of milk to yoghurt involves a specific group of bacteria called the lactic acid bacteria (*Lactobacillus* species). This is a general name attributed to those bacteria, which produce lactic acid as they grow. Acidic foods are less susceptible to spoilage than neutral or alkaline foods and hence the acid helps to preserve the product. Fermentations also result in a change in texture. In the case of milk, the acid causes the precipitation of milk protein to a solid curd.

2.2.4 Enzymes

The changes that occur during fermentation of foods are the result of enzymic activity. Enzymes are complex proteins produced by living cells to carry out specific biochemical reactions. They are known as catalysts since their role is to initiate and control reactions, rather than being used in a reaction. Because they are proteinaceous in nature, they are sensitive to fluctuations in temperature, pH, moisture content, ionic strength and concentrations of substrate and inhibitors. Each enzyme has requirements at which it will operate most efficiently. Extremes of temperature and pH will denature the protein and destroy enzyme activity. Because they are so sensitive, enzymic reactions can easily be controlled by slight adjustments to temperature, pH or other reaction conditions. In the food industry, enzymes have several roles - the liquefaction and saccharification of starch, the conversion of sugars and the modification of proteins. Microbial enzymes play a role in the fermentation of fruits and vegetables.

Nearly all food fermentations are the result of more than one microorganism, either working together or in a sequence. For example, vinegar production is a joint effort between yeast and acetic acid forming bacteria. The yeast convert sugars to alcohol, which is the substrate, required by the *acetobacter* to produce acetic acid. Bacteria from different species and the various micro-organisms - yeast and moulds -all have their own preferences for growing conditions, which are set within narrow limits. There are very few pure culture fermentations. An organism that initiates fermentation will grow there until its by-products inhibit further growth and activity. During this initial growth period, other organisms develop which are ready to take over when the conditions become intolerable for the former ones.

In general, growth will be initiated by bacteria, followed by yeasts and then moulds. There are definite reasons for this type of sequence. The smaller microorganisms are the ones that multiply and take up nutrients from the surrounding area most rapidly. Bacteria are the smallest of microorganisms, followed by yeasts and moulds. The smaller bacteria, such as *Leuconostoc* and *Streptococcus* grow and ferment more rapidly than their close relations and are therefore often the first species to colonise a substrate (Mountney and Gould, 1988)[43].

Table 2.2 Microorganisms commonly found in fermenting fruit and vegetables

Organism	Type	Optimum conditions	Reactions
Acetobacter genus	Aerobic rods	$a_w >=0.9$	Oxidise organic compounds (alcohol) to organic acids (acetic acid). Important in vinegar production.
A. aceti A. pasteurianus A. peroxydans			
Streptococcaceae Family	Gram positive cocci	Acid tolerant $a_w >=0.9$	
Streptococcus genus S. faecalis S. bovis S. thermophilus			Homofermentative. Most common in dairy fermentations, but S. Faecalis is common in vegetable products. Tolerate salt and can grow in high pH media.
Leuconostoc genus L. mesenteroides L. dextranicum L. paramesenteroides L. oenos	Gram positive cocci		Heterofermentative. Produce lactic acid, plus acetic acid, ethanol and carbon dioxide from glucose. Small bacteria, therefore have an important role in initiating fermentations. L. oenos is often present in wine. It can utilise malic acid and other organic acids.
Pediococcus genus P. cerevisiae P. acidilactici P. pentosaceus			Saprophytic organisms found in fermenting vegetables, mashes, beer and wort. Produce inactive lactic acid.
Lactobacillaceae Family	Gram positive rods. Non-motile	Acid tolerant $a_w >=0.9$	metabolise sugars to lactic acid, acetic acid, ethyl alcohol and carbon dioxide.
Lactobacillus genus			The genus is split into two types – homo- and hetero-fermenters. Saprophytic organisms. Produce greater amounts of acid than the cocci
Homofermentative Lactobacillus spp. L. delbrueckii L. leichmannii L. plantarum L. lactis L. acidophilus			Produce only lactic acid. L. plantarum important in fruit and vegetable fermentation. Tolerates high salt concentration.
Heterofermentative Spp. L. brevis L. fermentum L. buchneri			Produce lactic acid (50%) plus acetic acid (25%), ethyl alcohol and carbon dioxide (25%). L. brevis is the most common. Widely distributed in plants and animals. Partially reduces fructose to mannitol.
Yeasts		Tolerate acid, 40% sugar $a_w >=0.85$	
Saccharomyces Cerevisiae S. pombe	Many aerobic, some anaerobes	pH 4-4.5 20-30°C	S. cerevisiae can shift its metabolism from a fermentative to an oxidative pathway, depending on oxygen availability. Most yeasts produce alcohol and carbon dioxide from sugars.
Debaromyces			Tolerant of high salt concentrations
Zygosaccharomyces rouxii			Tolerates high salt concentration and low a_w
Candida species Geotrichum candidum			

2.3 Desirable fermentations

It is essential with any fermentation to ensure that only the desired bacteria, yeasts or moulds start to multiply and grow on the substrate. This has the effect of suppressing other microorganisms which may be either pathogenic and cause food poisoning or will generally spoil the fermentation process, resulting in an end product which is neither expected nor desired. An everyday example used to illustrate this point is the differences in spoilage between pasteurised and unpasteurised milk. Unpasteurised milk will spoil naturally to produce a sour tasting product, which can be used in baking to improve the texture of certain breads. Pasteurised milk, however, spoils (non-desirable fermentation) to produce an unpleasant product, which has to be disposed of. The reason for this difference is that pasteurisation (despite being a very important process to destroy pathogenic microorganisms) changes the microorganism environment and if pasteurised milk is kept unrefrigerated for some time, undesirable microorganisms start to grow and multiply before the desirable ones. In the case of unpasteurised milk, the non-pathogenic lactic acid bacteria start to grow and multiply at a greater rate that any pathogenic bacteria. Not only do the larger numbers of lactic acid bacteria compete more successfully for the available nutrients, but also as they grow they produce lactic acid, which increases the acidity of the substrate and further suppresses the bacteria, which cannot tolerate an acid environment.

Most food spoilage organisms cannot survive in either alcoholic or acidic environments. Therefore, the production of both these end products can prevent a food from spoilage and extend the shelf life. Alcoholic and acidic fermentations generally offer cost effective methods of preserving food for people in developing countries, where more sophisticated means of preservation are unaffordable and therefore not an option.

The principles of microbial action are identical both in the use of microorganisms in food preservation, such as through desirable fermentations, and also as agents of destruction via food spoilage. The type of organisms present and the environmental conditions will determine the nature of the reaction and the ultimate products. By manipulating the external reaction conditions, microbial reactions can be controlled to produce desirable results. There are several means of altering the reaction environment to encourage the growth of desirable organisms. These are discussed below.

2.4 Manipulation of microbial growth and activity

There are six major factors that influence the growth and activity of microorganisms in foods. These are moisture, oxygen concentration, temperature, nutrients, pH and inhibitors (Mountney and Gould, 1988)[44]. The food supply available to the microorganisms depends on the composition of the food on which they grow. All microorganisms differ in their ability to metabolise proteins, carbohydrates and fats. Obviously, by manipulating any of these six factors, the activity of microorganisms within foods can be controlled.

2.4.1 Moisture

Water is essential for the growth and metabolism of all cells. If it is reduced or removed, cellular activity is decreased. For example, the removal of water from cells by drying or the change in state of water (from liquid to solid) affected by freezing, reduces the availability of water to cells (including microbial cells) for metabolic activity. The form in which water exists within the food is important as far as microbial activity is concerned. There are two types of water - free and bound. Bound water is present within the tissue and is vital to all the physiological processes within the cell. Free water exists in and around the tissues and can be removed from cells without seriously interfering with the vital processes. Free water is essential for the survival and activity of microorganisms. Therefore, by removing free water, the level of microbial activity can be controlled. The amount of water available for microorganisms is referred to as the water activity (a_w). Pure water has a water activity of 1.0. Bacteria require more water than yeasts, which require more water than moulds to carry out their metabolic activities. Almost all microbial activity is inhibited below a_w of 0.6. Most fungi are inhibited below a_w of 0.7, most yeasts are inhibited below a_w of 0.8 and most bacteria below a_w 0.9. Naturally, there are exceptions to these guidelines and several species of microorganism can exist outside the stated range. See table for further information on water activity and microbial action. Altering the amount of free water available can change the water activity of foods. There are several ways to achieve this – drying to remove water; freezing to change the state of water from liquid to solid; increasing or decreasing the concentration of solutes by adding salt or sugar or other hydrophilic compounds (salt and sugar are the two common additives used for food preservation). Addition of salt or sugar to a food will bind free water and so decrease the a_w. Alternatively, decreasing the concentration will increase the amount of free water and in turn the a_w. Manipulation of the a_w in this manner can be used to encourage the growth of favourable microorganisms and discourage the growth of spoilage ones.

Table 2.3 Water activity for microbial reactions

A_w	Phenomenon	Examples
1.00		Highly perishable foods
0.95	Pseudomonas, Bacillus, Clostridium perfringens and some yeasts inhibited	Foods with 40% sucrose or 7% salt
0.90	Lower limit for bacterial growth. Salmonella, Vibrio parahaemolyticus, Clostridium botulinum, Lactobacillus and some yeasts and fungi inhibited	Foods with 55% sucrose, 12% salt. Intermediate-moisture foods (a_w = 0.90-0.55)
0.85	Many yeasts inhibited	Foods with 65% sucrose, 15% salt
0.80	Lower limit for most enzyme activity and growth of most fungi. Staphylococcus aureus inhibited	Fruit syrups
0.75	Lower limit for halophilic bacteria	Fruit jams
0.70	Lower limit for growth of most xerophilic fungi	
0.65	Maximum velocity of Maillard reactions	
0.60	Lower limt for growth of osmophilic or xerophilic yeasts and fungi	Dried fruits (15-20% water)
0.55	Deoxyribose nucleic acid (DNA) becomes disordered (lower limit for life to continue)	
0.50		Dried foods (a_w=0-0.55)
0.40	Maximum oxidation velocity	
0.25	Maximum heat resistance of bacterial spores	

Taken from Fellows (1988)[45].

2.4.2 Oxidation-Reduction potential

Oxygen is essential to carry out metabolic activities that support all forms of life. Free atmospheric oxygen is utilised by some groups of microorganisms, while others are able to metabolise the oxygen, which is bound to other compounds such as carbohydrates. This bound oxygen is in a reduced form.

Microorganisms can be broadly classified into two groups - aerobic and anaerobic. Aerobes grow in the presence of atmospheric oxygen while anaerobes grow in the absence of atmospheric oxygen. In the middle of these two extremes are the facultative anaerobes which can adapt to the prevailing conditions and grow in either the absence or presence of atmospheric oxygen. Microaerophilic organisms grow in the presence of reduced amounts of atmospheric oxygen. Thus, controlling the availability of free oxygen is one means of controlling microbial activity within a food. In aerobic fermentations, the amount of oxygen present is one of the limiting factors. It determines the type and amount of biological product obtained the amount of substrate consumed and the energy released from the reaction.

Moulds do not grow well in anaerobic conditions; therefore they are not important in terms of food spoilage or beneficial fermentation, in conditions of low oxygen availability.

2.4.3 Temperature

Temperature affects the growth and activity of all living cells. At high temperatures, organisms are destroyed, while at low temperatures, their rate of activity is decreased or suspended. Microorganisms can be classified into three distinct categories according to their temperature preference (see table2.4).

Table 2.4 Classification of bacteria according to temperature requirements.

	Temperature required for growth ^{0}C			
Type of bacteria	Minimum	optimum	maximum	General sources of bacteria
Psychrophilic	0 to 5	15 to 20	30	Water and frozen foods
Mesophilic	10 to 25	30 to 40	35 to 50	Pathogenic and non-pathogenic bacteria
Thermophilic	25 to 45	50 to 55	70 to 90	Spore forming bacteria from soil and water

(Taken from Mountney and Gould, (1988)[46].

2.4.4 Nutritional requirements

The majority of organisms are dependent on nutrients for both energy and growth. Organisms vary in their specificity towards different substrates and usually only colonise foods which contain the substrates they require. Sources of energy vary from simple sugars to complex carbohydrates and proteins. The energy requirements of microorganisms are very high. Limiting the amount of substrate available can check their growth.

2.4.5 Hydrogen ion concentration (pH)

The pH of a substrate is a measure of the hydrogen ion concentration. A food with a pH of 4.6 or less is termed a high acid or acid food and will not permit the growth of bacterial spores. Foods with a pH above 4.6. are termed low acid and will not inhibit the growth of bacterial spores. By acidifying foods and achieving a final pH of less than 4.6, most foods are resistant to bacterial spoilage.

The optimum pH for most microorganisms is near the neutral point (pH 7.0). Certain bacteria are acid tolerant and will survive at reduced pH levels. Notable acid-tolerant bacteria include *the Lactobacillus* and *Streptococcus* species, which play a role in the fermentation of dairy and vegetable products. Moulds and yeasts are usually acid tolerant and are therefore associated with spoilage of acidic foods.

Microorganisms vary in their optimal pH requirements for growth. Most bacteria favour conditions with a near neutral pH (7). Yeasts can grow in a pH range of 4 to 4.5 and moulds can grow from pH 2 to 8.5, but favour an acid pH. The varied pH requirements of different groups of micro-organisms is used to good effect in fermented foods where successions of micro-organisms take over from each other as the pH of the environment changes. For instance, some groups of microorganisms ferment sugars so that the pH becomes too low for the survival of those microbes. The acidophilic microorganisms then take over and continue the reaction. The affinity for different pH can also be used to good effect to occlude spoilage organisms.

2.4.6 Inhibitors

Many chemical compounds can inhibit the growth and activity of microorganisms. They do so by preventing metabolism, denaturation of the protein or by causing physical damage to the cell. The production of substrates as part of the metabolic reaction also acts to inhibit microbial action.

2.5 Controlled fermentations

Controlled fermentations are used to produce a range of fermented foods, including sauerkraut, pickles, olives, vinegar, dairy and other products. Controlled fermentation is a form of food preservation since it generally results in a reduction of acidity of the food, thus preventing the growth of spoilage micro-organisms. The two most common acids produced are lactic acid and acetic acid, although small amounts of other acids such as propionic, fumaric and malic acid are also formed during fermentation.

It is highly probable that the first controlled food fermentations came into existence through trial and error and a need to preserve foods for consumption later in the season. It is possible that the initial attempts at preservation involved the addition of salt or seawater. During the removal of the salt prior to consumption, the foods would pass through stages favourable to acid fermentation. Although the process worked, it is likely that the causative agents were unknown. Today, there are numerous examples of controlled fermentation for the preservation and processing of foods. However, only a few of these have been studied in any detail - these include sauerkraut, pickles, kimchi, beer, wine and vinegar production. Although the general principles and processes for many of the fermented fruit and vegetable products are the same -relying mainly on lactic acid and acetic acid forming bacteria, yeasts and moulds, the reactions have not been quantified for each product. The reactions are usually very complex and involve a series of microorganisms, either working together or in succession to achieve the final product.

CHAPTER 3

YEAST FERMENTATIONS

3.1 What are yeasts?

A yeast is a unicellular fungus which reproduces asexually by budding or division, especially the genus *Saccharomyces* which is important in food fermentations (Walker, 1988)[47]. Yeasts and yeast-like fungi are widely distributed in nature. They are present in orchards and vineyards, in the air, the soil and the intestinal tract of animals. Like bacteria and moulds, they can have beneficial and non-beneficial effects in foods. Most yeasts are larger than most bacteria. The most well known examples of yeast fermentation are in the production of alcoholic drinks and the leavening of bread. For their participation in these two processes, yeasts are of major importance in the food industry.

Some yeasts are chromogenic and produce a variety of pigments, including green, yellow and black. Others are capable of synthesising essential B group vitamins.

Although there is a large diversity of yeasts and yeast-like fungi, (about 500 species), only a few are commonly associated with the production of fermented foods. They are all either ascomycetous yeasts or members of the genus *Candida.* Varieties of the *Saccharomyces cervisiae* genus are the most common yeasts in fermented foods and beverages based on fruit and vegetables. All strains of this genus ferment glucose and many ferment other plant-derived carbohydrates such as sucrose, maltose and raffinose. In the tropics, *Saccharomyces pombe* is the dominant yeast in the production of traditional fermented beverages; especially those derived from maize and millet (Adams and Moss, 1995)[48].

3.2 Conditions necessary for fermentation

Most yeasts require an abundance of oxygen for growth, therefore by controlling the supply of oxygen, their growth can be checked. In addition to oxygen, they require a basic substrate such as sugar. Some yeasts can ferment sugars to alcohol and carbon dioxide in the absence of air but require oxygen for growth. They produce ethyl alcohol and carbon dioxide from simple sugars such as glucose and fructose.

$C_6H_{12}O_6$	$\Rightarrow$	$2C_2H_5OH$	+	$2CO_2$
Glucose	yeast	ethyl alcohol	+	carbon dioxide

In conditions of excess oxygen (and in the presence of acetobacter) the alcohol can be oxidised to form acetic acid. This is undesirable if the end product is a fruit alcohol, but is a technique employed for the production of fruit vinegars (see later section on mixed fermentations).

Yeasts are active in a very broad temperature range - from 0 to 50° C, with an optimum temperature range of 20° to 30° C.

The optimum pH for most microorganisms is near the neutral point (pH 7.0). Moulds and yeasts are usually acid tolerant and are therefore associated with the spoilage of acidic foods. Yeasts can grow in a pH range of 4 to 4.5 and moulds can grow from pH 2 to 8.5, but favour an acid pH (Mountney and Gould, 1988)[49].

In terms of water requirements, yeasts are intermediate between bacteria and moulds. Bacteria have the highest demands for water, while moulds have the least need. Normal yeasts require a minimum water activity of 0.85 or a relative humidity of 88%.

Yeasts are fairly tolerant of high concentrations of sugar and grow well in solutions containing 40% sugar. At concentrations higher than this, only a certain group of yeasts – the osmophilic type – can survive. There are only a few yeasts that can tolerate sugar concentrations of 65-70% and these grow very slowly in these conditions (Board, 1983)[50]. Some yeasts – for example the *Debaromyces* - can tolerate high salt concentrations. Another group which can tolerate high salt concentrations and low water activity is *Zygosaccharomyces rouxii*, which is associated with fermentations in which salting is an integral part of the process.

3.3 Production of fruit alcohol

Alcohol and acids are two primary products of fermentation; both used to good effect in the preservation of foods. Several alcohol-fermented foods are preceded by an acid fermentation and in the presence of oxygen and *acetobacter*, alcohol can be fermented to produce acetic acid. Most food spoilage organisms cannot survive in either alcoholic or acidic environments. Therefore, the production of both these end products can prevent a food from undergoing spoilage and extend its shelf life.

Primitive wines and beers have been produced, with the aid of yeasts, for thousands of years; although it was not until about four hundred years ago that microorganisms associated with the fermentation were observed and identified. It was not until the 1850's that Louis Pasteur demonstrated unequivocally the involvement of yeasts in the production of wines and beers (Fleet, 1998)[51]. Since then, the knowledge of yeasts and the conditions necessary for fermentation of wine and beer has increased to the point where pure culture fermentations are now used to ensure consistent product quality. Originally, alcoholic fermentations would have been spontaneous events that resulted from the activity of microorganisms naturally present. These non-scientific methods are still used today for the home preparation of many of the world's traditional beers and wines.

Alcoholic drinks fall into two broad categories: wines and beers. Wines are made from the juice of fruits and beers from cereal grains. The principal carbohydrates in fruit juices are soluble sugars; the principal carbohydrate in grains is starch, an insoluble polysaccharide. The yeasts that bring about alcoholic fermentation can attack soluble sugars but do not produce starch-splitting enzymes. Wines can therefore be made by the direct fermentation of the raw material, while the production of beer requires the hydrolysis of starch to yield sugars fermentable by yeast, as a preliminary step (Stanier, Dourdoff and Adelberg, 1972)[52].

Raw fruit juice is usually a strongly acidic solution, containing from 10 to 25 percent soluble sugars. Its acidity and high sugar concentration make it an unfavourable medium for the growth of bacteria but highly suitable for yeasts and moulds. Raw fruit juice naturally contains many yeasts, moulds, and bacteria, derived from the surface of the fruit. Normally the yeast used in alcoholic fermentation is a strain of the species *Saccharomyces cerevisiae* (Adams, 1985)[53].

The fermentation may be allowed to proceed spontaneously, or can be "started" by inoculation with a must that has been previously successfully fermented by *S. cerevisiae* var. *ellipsoideus*. Many modern wineries eliminate the original microbial population of the must by pasteurisation or by treatment with sulphur dioxide. The must is then inoculated with a *starter culture* derived from a pure culture of a suitable strain of wine yeast. This procedure eliminates many of the uncertainties and difficulties of older methods. At the start of the fermentation, the must is aerated slightly to build up a large and vigorous yeast population; once fermentation sets in, the rapid production of carbon dioxide maintains anaerobic conditions, which prevent the growth of undesirable aerobic organisms, such as bacteria and moulds. The temperature of fermentation is usually from 25 to 30°C, and the duration of the fermentation process may extend from a few days to two weeks. As soon as the desired degree of sugar disappearance and alcohol production has been attained, the microbiological phase of wine making is over. Thereafter, the quality and stability of the wine depend very largely on preventing further microbial activity, both during the "ageing" in wooden casks and after bottling (Stanier *et al*, 1972)[54].

At all stages during its manufacture, fruit juice alcohol is subject to spoilage by undesirable microorganisms. Pasteur, whose descriptions of the organisms responsible and recommendations for overcoming them are still valid today, first scientifically explored the problem of the "diseases" of wines. The most serious aerobic spoilage processes are brought about by film-forming yeasts and acetic acid bacteria, both of which grow at the expense of the alcohol, converting it to acetic acid or to carbon dioxide and water. The chief danger from these organisms arises when access of air is not carefully regulated during ageing. Much more serious are the diseases caused by fermentative bacteria, particularly rod-shaped lactic acid bacteria, which utilise any residual sugar and impart a mousy taste to the wine. Such wines are known as turned wines. Since oxygen is unnecessary for the growth of lactic acid bacteria, wine spoilage of this kind can occur even after bottling. These risks of spoilage can be minimised by pasteurisation after bottling (Stanier *et al*, 1972)[55].

3.3.1 Grape wine

Grape wine is perhaps the most common fruit juice alcohol. Because of the commercialisation of the product for industry, the process has received most research attention and is documented in detail.

The production of grape wine involves the following basic steps: crushing the grapes to extract the juice; alcoholic fermentation; maltolactic fermentation if desired; bulk storage and maturation of the wine in a cellar; clarification and packaging. Although the process is fairly simple, quality control demands that the fermentation be carried out under controlled conditions to ensure a high quality product.

The distinctive flavour of grape wine originates from the grapes as raw material and subsequent processing operations. The grapes contribute trace elements of many volatile substances (mainly terpenes) which give the final product the distinctive fruity character. In addition, they contribute non-volatile compounds (tartaric and malic acids) which impact on flavour and tannins, which give bitterness and astringency. The latter are more prominent in red wines as the tannin components are located in the grape skins.

Although yeasts are the principal organisms involved, filamentous fungi, lactic acid bacteria, acetic acid bacteria and other bacterial groups all play a role in the production of alcoholic fruit products (Fleet, 1998)[56].

Normal grapes harbour a diverse micro-flora, of which the principal yeasts (*Saccharomyces cerevisiae*) involved in desirable fermentation are in the minority. Lactic acid bacteria and acetic acid bacteria are also present. The proportions of each and total numbers present are dependent upon a number of external environmental factors including the temperature, humidity, stage of maturity, damage at harvest and application of fungicides. It is essential to ensure proliferation of the desired species at the expense of the non-desired ones. This is achieved through ensuring fermentation conditions are such to encourage *Saccharomyces* species.

The fermentation may be initiated using a starter culture of *Saccharomyces cerevisiae* – in which case the juice is inoculated with populations of yeast of 10^6 to 10^7 cfu[1]/ml juice. This approach produces a wine of generally expected taste and quality. If the fermentation is allowed to proceed naturally, utilising the yeasts present on the surface of the fruits, the end result is less controllable, but produces wines with a range of flavour characteristics. It is likely that natural fermentations are practised widely around the world, especially for home production of wine.

During alcoholic fermentation, yeasts are the prominent species. The composition of fruit juice – its acid and sugar level and low pH favour the growth of yeasts and production of ethanol that restricts the growth of bacteria and fungi.

In natural fermentations, there is a progressive pattern of yeast growth. Several species of yeast, including *Kloeckera, Hanseniaspora, Candida* and *Metschnikowia*, are active for the first two to three days of fermentation. The build up of end products (ethanol) is toxic to these yeasts and they die off, leaving *Saccharomyces cerevisiae* to continue the fermentation to the end. *S. cerevisiae* can tolerate much higher levels of ethanol (up to 15% v/v or more) than the other species who only tolerate up to 5 or 8% alcohol (Fleet, 1998)[57]. Because of its tolerance of alcohol, *S. cerevisiae* dominates wine fermentation and is the species that has been commercialised for starter cultures.

Traditionally, fermentation was carried out in large wooden barrels or concrete tanks. Modern wineries now use stainless steel tanks, as these are more hygienic and provide better temperature control. White wines are fermented at 10 to 18° C for about seven to fourteen days. The low temperature and slow fermentation favours the retention of volatile compounds. Red wines are fermented at 20 to 30°C for about seven days. This higher temperature is necessary to extract the pigment from the grape skins (Fleet, 1998)[58].

---

[1] cfu refers to coliform forming units. It is a measure of the activity of microorganisms.

3.3.2 Factors affecting wine fermentation.

There are several variables, which can affect the fermentation process and final quality of wine. Factors, which are most important to control, are:

- the clarification and pre-treatment of juice
- chemical composition of the juice
- temperature of the fermentation
- the influences of other microorganisms.

Clarification and pre-treatment of juice

Excessive clarification removes many of the natural yeasts and flora. This is beneficial if a tightly controlled induced fermentation is desired, but less so if the fermentation is a natural one. Long periods of settling out however, encourage the growth of natural flora, which can contribute to the fermentation.

Chemical composition of juice

The main constituents of grape juice are glucose (75 to 150 g/l), fructose (75 to 150 g/l), tartaric acid (2 to 10 g/l), malic acid (1 to 8 g/l) and free amino acids (0.2 to 2.5 g/l). The main reaction is the fermentation of glucose and fructose to ethanol and carbon dioxide. However, the presence of nitrogenous and sulphurous products also contributes to the fermentation. The addition of sulphur dioxide to the juice delays the growth of yeast, but does not necessarily inhibit growth of the non-Saccharomyces strains. Fruits generally contain sufficient substrates - soluble sugars - for the yeast to ferment and convert into an acceptable concentration of alcohol. Sugar can be added to fruit juices with a low sugar content, to increase the amount of fermentable substrate.

Temperature

Temperature has an impact on the growth and activity of different strains of yeast. At temperatures of 10 to 15°C, the non-Saccharomyces species have an increased tolerance to alcohol and therefore have the potential to contribute to the fermentation.

Influence of other micro-organisms

Other microorganisms have the potential to influence wine production at all stages of the process. Prior to harvest, yeasts grow on the surface of grapes. Fungicides are used in an attempt to control their growth, but these disturb the natural balance of flora, thus making it difficult to carry out a 'natural' fermentation. Overuse of fungicides can lead to the development of resistant strains of yeast, which have the potential to produce toxins, which destroy the desirable yeast species. These yeasts are known as 'killer' strains. Other microbes have further chances to influence the fermentation during the clarification process, after fermentation and during maturation and bottling when *acetobacter* species can oxidise the alcohol and produce acetic acid.

About two to three weeks after the alcoholic fermentation is finished wines often undergo a malo-lactic fermentation. This occurs naturally and lasts for about four weeks. It is a lactic acid fermentation, initiated by lactic acid bacteria resident in the wine. Inoculating the fermented wine with cultures of *Leuconostoc oenos* can start the process if it is desired. The main reaction of these bacteria is the decarboxylation of L-malic acid to L-lactic acid, which decreases the acidity of the wine and increases its pH by about 0.3 to 0.5 units. Wines produced from grapes grown in colder climates tend to have a higher concentration of malic acid and a lower pH (3.0 to 3.5) and the taste benefits from this slight decrease in acidity. The benefits of this process are that it imparts a more mellow flavour to the wine. The growth of malo-lactic bacteria also contributes to the taste of the wine. Wines that have undergone a malo-lactic fermentation appear to be less susceptible to any further damage from other bacteria. This could be because *L. oenos* has used up all available substrate, or it may have secreted bacteriocins which prevent the growth of other species (Fleet, 1998)[59]. Although the malo-lactic fermentation seems to be a useful process, not all wines benefit from it. Wines produced from grapes in warmer climates tend to be less acidic (pH > 3.5) and a further reduction in acidity may have adverse effects on the quality of the wine. Decreasing the acidity also increases the pH to values which can allow spoilage organisms to multiply. It is difficult to prevent the malo-lactic fermentation from taking place naturally, especially later on after the wine has been bottled. In low acid wines, the acidity may be adjusted after this fermentation has taken place. The malo-lactic fermentation can be prevented by controlling several factors: the wine pH (< 3.2); ethanol content (> 14%) and levels of sulphur dioxide (>50 mg/l). The bacteriocin nisin can also be used to control the growth of malo-lactic bacteria. However the subtle blend of aromas and flavours that contribute to the final taste may be lost by such stringent control.

The conversion of malic acid to lactic acid is one of the main reactions carried out by wine lactic acid bacteria. *L oenos* needs to be present in significant numbers (greater than 10^6 cfu/ml) for the reaction to take place at a suitable pace. The bacteria use residual pentose and hexose sugars in the wine as a substrate for growth. The main reaction is the deacidification (or decarboxylation) of malic acid. In addition to this, the by-products of the reaction impart flavours and aromas to the wine.

During storage, wines are prone to non-desirable microbial changes. Yeasts, lactic acid bacteria, acetic acid bacteria and fungi can all spoil or taint wines after the fermentation process is completed. The changes that occur are increased acidification through the formation of acetic and other acids from alcohol; increased carbonation through a secondary fermentation of residual sugars and flavour changes through the metabolism of numerous compounds.

CHAPTER 4

PRODUCTS OF YEAST FERMENTATATION

The major products of yeast fermentation are alcoholic drinks and bread. With respect to fruits and vegetables, the most important products are fermented fruit juices and fermented plant saps. Virtually any fruit or sugary plant sap can be processed into an alcoholic beverage. The process is well known being essentially an alcoholic fermentation of sugars to yield alcohol and carbon dioxide.

It should be noted that alcohol production requires special licences or is prohibited in many countries.

4.1 Fermented fruit juices

There are many fermented drinks made from fruit in Africa, Asia and Latin America. These include drinks made from bananas, grapes and other fruit. Grape wine is perhaps the most economically important fruit juice alcohol. It is of major economic importance in Chile, Argentina, South Africa, Georgia, Morocco and Algeria. Because of the commercialisation of the product for industry, the process has received most research attention and is documented in detail. Banana beer is probably the most wide spread alcoholic fruit drink in Africa that is of cultural importance in certain areas. Alcoholic fruit drinks are made from many other fruits including dates in North Africa, pineapples in Latin America and jackfruits in Asia.

4.1.1 Red Grape wine

Location of production

Red grape wines are made in many African, Asian and Latin American countries including Algeria, Morocco and South Africa.

Product description

Red grape wine is an alcoholic fruit drink of between 10 and 14% alcoholic strength. The colour ranges from a light red to a deep dark red. It is made from the fruit of the grape plant (*Vitis vinifera*). There are many varieties of grapes used including Cabernet Sauvignon, Grenache, Nebula, Pinot Noir, and Torrents (Ranked, Kill and Baker, 1997)[60]. The skins of the grape are allowed to ferment in red wine production, to allow for the extraction of colour and tannins, which contribute to the flavour. The grapes contribute trace elements of many volatile substances, which give the final product the distinctive fruity character. In addition, they contribute non-volatile compounds (tartaric and malic acids) which impact on flavour and tannins, which give bitterness and astringency.

Raw material preparation

Ripe and undamaged grapes should be used. Red grapes are crushed to yield the juice plus skins, which is known as must.

Processing

The crushed grapes are transferred to fermentation vessels. The ethanol formed during this fermentation assists with the extraction of pigments from the skins. This takes between 24 hours and three weeks depending on the colour of the final product required.

The skins are then removed and the partially fermented wine is transferred to a separate tank to complete the fermentation. The fermentation can be from naturally occurring yeasts on the skin of the grape or using a starter culture of *Saccharomyces cerevisiae* – in which case the juice is inoculated with populations of yeast. This approach produces a wine of generally expected taste and quality. If the fermentation is allowed to proceed naturally, utilising the yeasts present on the surface of the fruits, the end result is less controllable, but produces wines with a range of flavour characteristics (Fleet, 1998)[61], (Rhodes and Fletcher, 1966)[62], (Colquichagua, 1994)[63].

Traditionally, fermentation was carried out in large wooden barrels or concrete tanks. Modern wineries now use stainless steel tanks, as these are more hygienic and provide better temperature control.

Fermentation stops naturally when all the fermentable sugars have been converted to alcohol or when the alcoholic strength reaches the limit of tolerance of the strain of yeast involved. Fermentation can be stopped artificially by adding alcohol, by sterile filtration or centrifugation (Ranken, Kill and Baker, 1997)[64].

Some wines can be drunk immediately. However most wines develop distinctive favours and aromas by ageing in wooden casks.

Flow diagram

Selection of grapes	Mature and undamaged grapes
⇓	
Crushing	Traditionally manually, but now by crushers
⇓	
Pre-fermentation	24 hours to three weeks depending on colour required
⇓	
Removal of skin	Can add sulphur dioxide to inhibit wild yeasts
⇓	
Fermentation	
⇓	
Maturation	Ageing to develop aromas and flavours

Packaging and storage

Traditionally wine was delivered to the point of sale in casks. The product is traditionally packaged in glass bottles with corks, made from the bark of the cork oak (*Quercus suber*). The bottles should be kept out of direct sunlight. During storage, wines are prone to non-desirable microbial changes. Yeasts, lactic acid bacteria, acetic acid bacteria and fungi can all spoil or taint wines after the fermentation process is completed.

4.1.2 White Grape wine

Location of production

White grape wines are made in many African, Asian and Latin American countries including Algeria, Morocco and South Africa.

Product description

White grape wine is an alcoholic fruit drink of between 10 and 14% alcoholic strength. It is prepared from the fruit of the grape plant (*Vitis vinifera*), and is pale yellow in colour. There are many varieties used including Airen, Chardonnay, Palomino, Sauvignon Blanc and Ugni Blanc (Ranken, Kill and Baker, 1997)[65]. The main difference between red and white wines is the early removal of grape skins in white wine production. The distinctive flavour of grape wine originates from the grapes as raw material and subsequent processing operations. The grapes contribute trace elements of many volatile substances (mainly terpenes) which give the final product the distinctive fruity character.

Preparation of raw materials

Ripe and undamaged grapes should be used. The grapes are crushed to yield the juice and the skins are removed and separated out. Sometimes the juice is clarified by allowing it to stand for 24 to 48 hours at 5 to 10° C, by filtering or centrifugation. Pectolytic enzymes may be added to accelerate the breakdown of cell wall tissue and to improve the clarity of juice. Excessive clarification removes many of the natural yeasts and flora. This is beneficial if a tightly controlled induced fermentation is desired, but less so if the fermentation is a natural one. Long periods of settling out however, encourage the growth of natural flora, which can contribute to the fermentation.

Processing

The clarified juice is transferred to a fermentation tank where fermentation either begins spontaneously or is induced by the addition of a starter culture. Traditionally, fermentation was carried out in large wooden barrels or concrete tanks. Modern wineries now use stainless steel tanks, as these are more hygienic and provide better temperature control. White wines are fermented at 10 to 18° C for about seven to fourteen days. The low temperature and slow fermentation favours the retention of volatile compounds (Fleet, 1998)[66].

The fermentation can be from naturally occurring yeasts on the skin of the grape or using a starter culture of *Saccharomyces cerevisiae*. This approach produces a wine of generally expected taste and quality. If the fermentation is allowed to proceed naturally, utilising the yeasts present on the surface of the fruits, the end result is less controllable, but produces wines having a range of flavour characteristics. It is likely that natural fermentations are practised widely around the world, especially for home production of wine.

During storage, wines are prone to non-desirable microbial changes. Yeasts, lactic acid bacteria, acetic acid bacteria and fungi can all spoil or taint wines after the fermentation process is completed.

Flow diagram

Selection of grapes	Mature and undamaged grapes
⇓	
Crushing	Traditionally manual but now usually by crushers
⇓	
Removal of skins	
⇓	
Clarification	By standing, filtration or centrifugation
⇓	
Fermentation	
⇓	
Ageing	

Packaging and storage

Traditionally wine was delivered to the point of sale in casks. The product is traditionally packaged in glass bottles with corks, made from the bark of the cork oak (*Quercus suber*). The bottles should be kept out of direct sunlight. During storage, wines are prone to non-desirable microbial changes. Yeasts, lactic acid bacteria, acetic acid bacteria and fungi can all spoil or taint wines after the fermentation process is completed.

4.1.3 Banana beer

Location of production

Throughout Africa

Product description

Banana beer is made from bananas, mixed with a cereal flour (often sorghum flour) and fermented to an orange, alcoholic beverage. It is sweet and slightly hazy with a shelf life of several days under correct storage conditions. There are many variations in how the beer is made. For instance *Urwaga* banana beer in Kenya is made from bananas and sorghum or millet and *Lubisi* is made from bananas and sorghum.

Preparation of raw materials

Ripe bananas (*Musa* spp.) are selected. The bananas should be peeled. If the peels cannot be removed by hand then the bananas are not sufficiently ripe.

Processing

The first step of the process is the extraction of banana juice. Extraction of a high yield of banana juice without excessive browning or contamination by spoilage microorganisms and proper filtration to produce a clear product is of great importance. Grass is used as an aid in obtaining clarified juice.

One volume of water is added to every three volumes of banana juice. This makes the total soluble solids low enough for the yeast to act. Cereals are ground and roasted and added to improve the colour and flavour of the final product. The mixture is placed in a container, which is covered in polythene to ferment for 18 to 24 hours. The raw materials are not sterilised by boiling and therefore provide an excellent substrate for microbial growth. It is essential that proper hygienic procedures are followed and that all equipment is thoroughly sterilised to prevent contaminating bacteria from competing with the yeast and producing acid instead of alcohol. This can be done by cleaning with boiling water or with chlorine solution. Care is necessary to wash the equipment free of residual chlorine, as this would interfere with the actions of the yeast. Strict personal hygiene is also essential (Fellows, 1997)[67].

For many traditional fermented products, the microorganisms responsible for the fermentation are unknown to scientists. However there has been research to identify the microorganisms involved in banana beer production. The main microorganism involved, is *Saccharomyces cerevisiae* which is the same organism involved in the production of grape wine. However many other microorganisms associated with the fermentation have been identified. These varied according to the region of production (Davies, 1994)[68].

After fermentation the product is filtered through cotton cloth.

Flow diagram

Raw materials	Ripe bananas
⇓	
Peel	Peel by hand
⇓	
Remove residue	Use grass to knead or squeeze out the juice
⇓	
Mix with water	The water: banana juice ratio should be 1:3
⇓	
Mix with cereals	Mix with ground and roasted cereals to local taste
⇓	
Ferment	In plastic container. Leave to ferment for 18 to 24 hours.
⇓	
Filter	Through cotton cloth
⇓	
Pack	
⇓	
Store	

Packaging and storage

Packaging is usually only required to keep the product for its relatively short shelf life. Clean glass or plastic bottles are used. The product is kept in a cool place away from direct sunlight.

4.1.4 Cashew wine

Location of Production

Cashew wine is made in many countries in Asia and Latin America.

Product description

Cashew wine is a light yellow alcoholic drink prepared from the fruit of the cashew tree (*Ancardium occidentale*). It contains an alcohol content of between 6 and 12% alcohol.

Preparation of raw materials

In gathering the fruits and transporting them to the workshop, the prime purpose should be to have the fruit arrive in the very best condition possible. Cashew apples are sorted and only mature undamaged cashew apples should be selected. These should be washed in clean water.

Processing

The cashew apples are cut into slices to ensure a rapid rate of juice extraction when crushed in a juice press. The fruit juice is sterilised in stainless steel pans at a temperature of 85°C in order to eliminate wild yeast. The juice is filtered and treated either sodium or potassium metabisulphite to destroy or inhibit the growth of any undesirable types of microorganisms - acetic acid bacteria, wild yeasts and moulds.

Wine yeast (*Saccharomyees cerevisiae - var. ellipsoideus*) is added. Once the yeast is added, the contents are stirred well and allowed to ferment for about two weeks.

The wine is separated from the sediment. It is clarified by using fining agents such as gelatin, pectin or casein, which are mixed with the wine. Filtration is carried out with filter-aids such as fullers earth. The filtered wine is transferred to wooden vats.

The wine is then pasteurised at 50° - 60°C. Temperature should be controlled, so as not to heat it to about 70°C, since its alcohol content would vaporise at a temperature of 75°-78°C. It is then stored in wooden vats and subjected to ageing. At least six months should be allowed for ageing.

If necessary, wine is again clarified prior to bottling. During ageing, and subsequent maturing in bottles many reactions, including oxidation, occur with the formation of traces of esters and aldehydes, which together with the tannin and acids already present enhance the taste, aroma and preservative properties of the wine (Wimalsiri, Sinnatamby, Samaranayake and Samarsinghe, 1971)[69].

Flow diagram

Selection	Mature, sound cashew apples
⇓	
Slicing	To increase extraction of juice
⇓	
Crushing	
⇓	
Sterilisation	At 85°C
⇓	
Filtered	
⇓	
Inoculation	With *S. cerevisiae*
⇓	
Fermentation	Two weeks
⇓	
Filtered	
⇓	
Pasteurisation	50-60°C
⇓	
Ageing.	In wooden vats for 6 months

Packaging and storage

The product is packaged in glass bottles with corks. The bottles should be kept out of direct sunlight.

4.1.5 *Tepache*

Tepache is a light, refreshing beverage prepared and consumed throughout Mexico. In the past, *tepache* was prepared from maize, but nowadays various fruits such as pineapple, apple and orange are used. The pulp and juice of the fruit are allowed to ferment for one or two days in water with some added brown sugar. The mixture is contained in a lidless wooden barrel called a *'tepachera'*, which is covered with cheesecloth. After a day or two, the *tepache* is a sweet and refreshing beverage. If fermentation is allowed to proceed longer, it turns into an alcoholic beverage and later into vinegar. The microorganisms associated with the product include *Bacillus subtilis, B. graveolus* and the yeasts, *Torulopsis insconspicna, Saccharomyces cerevisiae* and *Candida queretana* (Aidoo, 1986)[70].

4.1.6 *Colonche*

Colonche is a sweet, fizzy beverage produced in Mexico by fermenting the juice of the fruits of the prickly pear cacti - mainly *Opuntia* species. The procedure for preparing *colonche* is essentially the same as has been followed for centuries. The cactus fruits are peeled and crushed to obtain the juice, which is boiled for 2-3 hours. After cooling, the juice is allowed to ferment for a few days. Sometimes old *colonche* or *tibicos* may be added as a starter. *Tibicos* are gelatinous masses of yeasts and bacteria, grown in water with brown sugar. They are also used in the preparation of *tepache*.

4.1.7 Fortified grape wines

Fortified wines are made in the Republic of South Africa and North Africa. Fortified wines are made by adding spirits to wines, either during or after fermentation, with the result that the alcohol content of the wines is raised to around 20 percent, i.e. approximately double that of table wines (Rose, 1961)[71].

4.1.8 Date wine

Date wines are popular in Sudan and North Africa. They are made using a variety of methodologies. *Dakhai* is produced by placing dates in a clean earthenware pot. For every one volume of dates between two and four volumes of boiling water are added. This is allowed to cool and is then sealed for three days. More warm water is then added and the container sealed again for seven to ten days. Many variations of date wine exist: *El madfuna* is produced by burying the earthenware pots underground. *Benti merse* is produced from a mixture of sorghum and dates. *Nebit* is produced from date syrup (Dirar, 1992)[72].

4.1.9 Sparkling grape wine

Sparkling grape wines are made in the Republic of South Africa. Sparkling wines can be made in one of three ways. The cheapest method is to carbonate wines under pressure. Unfortunately, the sparkle of these wines quickly disappears, and the product is considered inferior to the sparkling wines produced by the traditional method of secondary fermentation. This involves adding a special strain of wine yeast (*S. cerevisiae* var. *ellipsoideus*) - champagne yeast - to wine that has been artificially sweetened. Carbon dioxide produced by fermentation of the added sugar gives the wine its sparkle. In the original champagne method, which is still widely used today, this secondary fermentation is carried out in strong bottles, capable of withstanding pressure but early in the nineteenth century a method of fermenting the wine in closed tanks was devised, this being considerably cheaper than using bottles (Rose, 1961)[73].

4.1.10 Jackfruit wine

Jackfruit wine is an alcoholic beverage made by ethnic groups in the eastern hilly areas of India. As its name suggests, it is produced from the pulp of jackfruit (*Artocarpus heterophyllus*). Ripe fruit is peeled and the skin discarded. The seeds are removed and the pulp soaked in water. Using bamboo baskets, the pulp is ground to extract the juice, which is collected in earthenware pots. A little water is added to the pots along with fermented wine inoculum from a previous fermentation. The pots are covered with banana leaves and allowed to ferment at 18 to 30°C for about one week. The liquid is then decanted and drunk. During fermentation, the pH of the wine reaches a value of 3.5 to 3.8, suggesting that an acidic fermentation takes place at the same time as the alçoholic fermentation. Final alcohol content is about 7 to 8% within a fortnight (Steinkraus, 1996)[74].

4.2 Fermented plant saps

Virtually any sugary plant sap can be processed into an alcoholic beverage. The process is well known being essentially an alcoholic fermentation of sugars to yield alcohol and carbon dioxide. Many alcoholic drinks are made from the juices of plants including coconut palm, oil palm, wild date palm, nipa palm, raphia palm and kithul palm.

4.2.1 Palm wine

Location of production

Palm 'wine' is an important alcoholic beverage in West Africa where it is consumed by more than 10 million people.

Product description

Palm wine can be consumed in a variety of flavours varying from sweet unfermented to sour fermented and vinegary alcoholic drinks. There are many variations and names including *emu* and *ogogoro* in Nigeria and *nsafufuo* in Ghana. It is produced from sugary palm saps. The most frequently tapped palms are raphia palms (*Raphia hookeri* or *R. vinifera*) and the oil palm (*Elaeis guineense*). Palm wine has been found to be nutritious. The fermentation process increases the levels of thiamin, riboflavin, pyridoxin and vitamin B_{12}. Like many African alcoholic beverages, palm wine has a very short shelf life. The product is not preserved for more than one day. After this time accumulation of an excessive amount of acetic acid makes it unacceptable to consumers. The bark of a tree (*Saccoglottis gabonensis*) may be added as a preservative. The alkaloid and phenolic compounds which are extracted into the wine have antimicrobial effect (Odunfa, 1985) [75].

Preparation of raw materials

Sap is collected by tapping the palm. Tapping is achieved by making an incision between the kernels and a gourd is tied around to collect the sap which is collected a day or two later. The fresh palm juice is a sweet, clear, colourless juice containing 10-12 percent sugar and is neutral. The quality of the final wines is determined mostly by the conditions used in the collection of the sap. Often the collecting gourd is not washed between collections and residual yeasts in the gourd quickly begin the fermentation.

Processing

The sap is not heated and the wine is an excellent substrate for microbial growth. It is therefore essential that proper hygienic collection procedures are followed to prevent contaminating bacteria from competing with the yeast and producing acid instead of alcohol (Fellows, 1997) [76].

Fermentation starts soon after the sap is collected and within an hour or two, the sap becomes reasonably high in alcohol (up to 4%). If allowed to continue to ferment for more than a day, the sap begins turning into vinegar, although the vinegary flavour is preferred by some. Organisms responsible include *S. cerevisiae*, and *Schizosaccharomyces pombe*, and the bacteria *Lactobacillus plantarum* and *L. mesenteroides*. There are reports that the yeasts and bacteria originate from the gourd, palm tree, and tapping implements. However the high sugar content of the juice would seem to selectively favour the growth of yeasts which might originate from the air. This is supported by the fact that fermentation also takes place in plastic containers. Within 24 hours the initial pH is reduced from 7.4-6.8 to 5.5 and the alcohol content ranges from 1.5 to 2.1 percent. Within 72 hours the alcohol levels increase from 4.5 to 5.2 percent and the pH is 4.0. Organic acids present are lactic acid, acetic acid and tartaric acid (Odunfa, 1985)[77].

The main control points are extraction of a high yield of palm sap without excessive contamination by spoilage microorganisms, and proper storage to allow natural fermentation to take place.

Flow diagram

Cut	Cut 10-15 cm from the top of the trunk
⇓	
Tapping	A gourd is fixed below the cut.
⇓	
Collection	The sap is collected each day
⇓	
Fermentation	Natural fermentation starts as soon as sap is collected
⇓	
Filter	Optional
⇓	
Bottling	Clean bottles should be used

Packaging and storage

Packaging is usually only required to keep the product for its relatively short shelf life. Clean glass or plastic bottles should be used. The product should be kept in a cool place away from direct sunlight.

4.2.2 Toddy

Location of production

Throughout Asia, particularly India and Sri Lanka.

Product description

Toddy is an alcoholic drink made by the fermentation of the sap from a coconut palm. It is white and sweet with a characteristic flavour. It is between 4 and 6% alcohol and has a shelf life of about 24 hours.

Preparation of raw materials

The sap is collected by slicing off the tip of an unopened flower. The sap oozes out and can be collected in a small pot tied underneath the flower.

Processing

The fermentation starts as soon as the sap collects in the pots on the palms, particularly if a small amount of toddy is left in the pots. The toddy is fully fermented in six to eight hours. The product is usually sold immediately due to its short shelf life (Fellows, 1997)[78].

Flow diagram

Tapping	A small pot is fixed below the cut.
⇓	
Collection	The sap is collected each day
⇓	
Fermentation	Natural fermentation starts as soon as sap is collected
⇓	
Packaging	Clean bottles should be used

Packaging and storage

Packaging is usually only required to keep the product for its relatively short shelf life. This is usually clean glass or plastic bottles. The product should be kept in a cool place away from direct sunlight.

4.2.3 *Pulque*

Location of production

Pulque is the national drink in Mexico, where, it is claimed, it originated with the early Aztecs. *Pulque* is a traditional beverage that now forms the basis of a national industry, together with the spirits mezcal and tequila that are obtained from it. Pulque plays an important role in the nutrition of low-income people in Mexico with B vitamins being present in nutritionally important levels.

Product description

Pulque is a milky, slightly foamy, acidic and somewhat viscous beverage. It is obtained by fermentation of aguamiel, which is the name given to the juices of various cacti, notably *Agave atrovirens* and *A. americana, which are often,* called the "Century plant" in English. The alcohol content on pulque varies between six and seven percent. The beverage obtained upon distilling pulque is called "Mezcal", and if manufactured in the Tequila region from a numbered distillery, it is referred to as "Tequila". The drink is often considered an aphrodisiac. The name Ticyaol is given to a good strain that makes one particularly virile. Pulque is frequently the potion of choice used by women during menstruation.

Preparation of raw materials

The juices are extracted from the plants when they are eight to ten years old and fermentation takes place spontaneously, although occasionally the juices are inoculated with a starter from previous fermentations.

Processing

The juice is allowed to ferment naturally through a mixed fermentation although yeast (*Saccharomyces carbajali)* is the main actor. *Lacto-bacillus plantarum* produces lactic acid and the viscosity of pulque is caused by the activity of two species of *Leuconostoc* which produce dextrans (Wood and Hodge)[79]. During fermentation of the juices of the plant, the soluble solids are reduced from between 25-30% to 6%; the pH falls from 7.4 to between 3.5 and 4.0; the sucrose content falls from 15% to 1% and vitamin levels are increased. For instance the vitamin content (milligrams of vitamins per 100g of product) increases from 5 to 29 for thiamine, 54 to 515 for niacin and 18 to 33 for riboflavin (Steinkraus, K.H. (1992) [80].

Packaging and storage

Packaging is only required to keep the product for its relatively short shelf life. Clean glass or plastic bottles should be used. The product should be kept in a cool place away from direct sunlight.

4.2.3 Ulanzi (Bamboo Wine)

Location of production

East and Southern Africa.

Product description

Ulanzi is a fermented bamboo sap obtained by tapping young bamboo shoots during the rainy season. It is a clear, whitish drink with a sweet and alcoholic flavour.

Preparation of raw materials

The bamboo shoots should be young in order to obtain a high yield of sap. The growing tip is removed and a container fixed in place to collect the sap. The container should be clean in order to prevent contamination of the fresh sap.

Processing

The raw material is an excellent substrate for microbial growth and fermentation begins immediately after collection. Fermentation takes between five and twelve hours depending on the strength of the final product desired.

Packaging and storage

Packaging is usually only required to keep the product for its relatively short shelf life.

4.2.4 *Basi* (Sugar cane wine)

Basi is a sugar cane wine made in the Philipppines by fermenting boiled, freshly extracted, sugar cane juice. A dried powdered starter is used to initiate the fermentation. The mixture is allowed to ferment for up to three months, and to age for up to one year. The final product is light brown in colour and has a sweet and a sour flavour. A similar product called *shoto sake* is made in Japan (Steinkraus, 1996)[81].

4.2.5 Muratina

Muratina is an alcoholic drink made from sugar cane and *muratina* fruit in Kenya. The fruit is cut in half, sun dried and boiled in water. The water is removed and the fruit is again sun dried. The fruit is added to a small amount of sugar cane juice and incubated in a warm place for 24 hours, after which it is removed and sun dried. The dried fruit is then added to a barrel of sugar cane juice which is allowed to ferment for between one and four days. The final product has a sour alcoholic taste (Steinkraus, 1996)[82].

CHAPTER 5

BACTERIAL FERMENTATIONS

5.1 What are bacteria

Bacteria are "a large group of unicellular or multi-cellular organisms lacking chlorophyll, with a simple nucleus, multiplying rapidly by simple fission, some species developing a highly resistant resting (spore) phase; some species reproduce sexually, and some are motile. In shape they are spherical, rodlike, spiral, or filamentous. They occur in air, water, soil, rotting organic material, animals and plants. Saprophytic forms are more numerous than parasites. A few forms are autotrophic" (Walker, 1988)[83].

There are several bacterial families present in foods, the majority of which are concerned with food spoilage. The important role of bacteria in the fermentation of foods is often overlooked.

5.2 Lactic Acid Bacteria

The lactic acid bacteria are a group of Gram positive bacteria, non-respiring, non-spore forming, cocci or rods, which produce lactic acid as the major end product of the fermentation of carbohydrates. They are the most important bacteria in desirable food fermentations, being responsible for the fermentation of sour dough bread, sorghum beer, all fermented milks, cassava (to produce *gari* and *fufu*) and most "pickled" (fermented) vegetables. Historically, bacteria from the genera *Lactobacillus, Leuconostoc, Pediococcus* and *Streptococcus* are the main species involved. Several more have been identified, but play a minor role in lactic fermentations. Axelsson (1998) recently reviewed lactic acid bacteria[84].

Lactic acid bacteria carry out their reactions - the conversion of carbohydrate to lactic acid plus carbon dioxide and other organic acids - without the need for oxygen. They are described as microaerophilic as they do not utilise oxygen. Because of this, the changes that they effect do not cause drastic changes in the composition of the food. Some of the family are homofermentative, that is they only produce lactic acid, while others are heterofermentative and produce lactic acid plus other volatile compounds and small amounts of alcohol. *Lactobacillus acidophilus, L. bulgaricus, L. plantarum, L. caret, L. pentoaceticus, L brevis* and *L. thermophilus* are examples of lactic acid-producing bacteria involved in food fermentations. All species of lactic acid bacteria have their own particular reactions and niches, but overall, *L. plantarum* – a homofermenter -produces high acidity in all vegetable fermentations and plays the major role. All lactic acid producers are non-motile gram positive rods that need complex carbohydrate substrates as a source of energy. The lactic acid they produce is effective in inhibiting the growth of other bacteria that may decompose or spoil the food. Because the whole group is referred to as 'lactic acid bacteria' it might appear that the reactions they carry out are very simple, with the production of one substrate. This is far from the truth. The lactic acid bacteria are a diverse group of organisms with a diverse metabolic capacity. This diversity makes them very adaptable to a range of conditions and is largely responsible for their success in acid food fermentations.

Despite their complexity, the whole basis of lactic acid fermentation centres on the ability of lactic acid bacteria to produce acid, which then inhibits the growth of other non-desirable organisms. All lactic acid producers are micro-aerophilic, that is they require small amounts of oxygen to function. Species of the genera *Streptococcus* and *Leuconostoc* produce the least acid. Next are the heterofermentative species of *Lactobacillus* which produce intermediate amounts of acid, followed by the *Pediococcus* and lastly the homofermenters of the *Lactobacillus* species, which produce the most acid. Homofermenters, convert sugars primarily to lactic acid, while heterofermenters produce about 50% lactic acid plus 25% acetic acid and ethyl alcohol and 25% carbon dioxide. These other compounds are important as they impart particular tastes and aromas to the final product. The heterofermentative lactobacilli produce mannitol and some species also produce dextran.

Leuconostoc mesenteroides is a bacterium associated with the sauerkraut and pickle fermentations. This organism initiates the desirable lactic acid fermentation in these products. It differs from other lactic acid species in that it can tolerate fairly high concentrations of salt and sugar (up to 50% sugar). *L. mesenteroides* initiates growth in vegetables more rapidly over a range of temperatures and salt concentrations than any other lactic acid bacteria. It produces carbon dioxide and acids which rapidly lower the pH and inhibit the development of undesirable microorganisms. The carbon dioxide produced replaces the oxygen, making the environment anaerobic and suitable for the growth of subsequent species of lactobacillus.. Removal of oxygen also helps to preserve the colour of vegetables and stabilises any ascorbic acid that is present.

Organisms from the gram-positive *Propionibacteriaceae* family are responsible for the flavour and texture of some fermented foods, especially Swiss cheese, where they are responsible for the formation of 'eyes' or holes in the cheese. These bacteria break down lactic acid into acetic and propionic acids and carbon dioxide.

Several other bacteria, for *instance Leuconostoc citrovorum L. Dextranicum, Streptococcus lactis, S. Cremis, & liquefaciens* and *Brevibacterium* species are important in the fermentation of dairy products. They are not discussed in detail in this manuscript.

5.2.1 Lactic acid fermentation

The lactic acid bacteria belong to two main groups – the homofermenters and the heterofermenters. The pathways of lactic acid production differ for the two. Homofermenters produce mainly lactic acid, via the glycolytic (Embden–Meyerhof) pathway). Heterofermenters produce lactic acid plus appreciable amounts of ethanol, acetate and carbon dioxide, via the 6-phosphoglucanate/phosphoketolase pathway. All lactic acid bacteria except leuconostocs, group III lactobacilli, oenococci and weissellas use the glycolytic pathway. Normal conditions required for this pathway are excess sugar and limited oxygen. Axelsson (1998) [85] gives an in-depth account of the biochemical pathways for both homo- and hetero-fermenters.

Homolactic fermentation

The fermentation of 1 mole of glucose yields two moles of lactic acid;

$$\underset{\text{Glucose}}{C_6H_{12}O_6} \Rightarrow \underset{\text{lactic acid}}{2\ CH_3CHOHCOOH}$$

Heterolactic fermentation

The fermentation of 1 mole of glucose yields 1 mole each of lactic acid, ethanol and carbon dioxide

$C_6H_{12}O_6$	⇒	$CH_3CHOHCOOH$	+	C_2H_5OH	+	CO_2
Glucose		lactic acid	+	ethanol	+ carbon dioxide	

Table 5.1 Major lactic acid bacteria in fermented plant products.

Homofermenter	Facultative homofermenter	Obligate heterofermenter
Enterococcus faecium	*Lactobacillus bavaricus*	*Lactobacillus brevis*
Enterococcus faecalis	*Lactobacillus casei*	*Lactobacillus buchneri*
Lactobacillus acidophilus	*Lactobacillus coryniformis*	*Lactobacillus cellobiosus*
Lactobacillus lactis	*Lactobacillus curvatus*	*Lactobacillus confusus*
Lactobacillus delbrueckii	*Lactobacillus plantarum*	*Lactobacillus coprophilus*
Lactobacillusleichmannii	*Lactobacillus sake*	*Lactobacillus fermentatum*
Lactobacillus salivarius		*Lactobacillus sanfrancisco*
Streptococcus bovis		*Leuconostoc dextranicum*
Streptococcus thermophilus		*Leuconostoc mesenteroides*
Pediococcus acidilactici		*Leuconostoc paramesenteroides*
Pedicoccus damnosus		
Pediococcus pentocacus		

From Beuchat (1995)[86].

5.3 Acetic acid bacteria

A second group of bacteria of importance in food fermentations are the acetic acid producers from the *Acetobacter* species. *Acetobacter* are important in the production of vinegar (acetic acid) from fruit juices and alcohols. The same reaction also occurs in wines, oxygen permitting, where the *acetobacter* can cause undesirable changes – the oxidation of alcohol to acetic acid. This produces a vinegary off-taste in the wine.

The most desirable action of acetic acid bacteria is in the production of vinegar. The vinegar process is essentially a two-stage process, where yeasts convert sugars into alcohol, followed by *acetobacter*, which oxidise alcohol to acetic acid. For this reason, the products of *acetobacter* fermentation are discussed in more detail in chapter 7 on mixed fermentations.

5.3.1 Acetic acid fermentation

Acetobacter convert alcohol to acetic acid in the presence of excess oxygen.

Oxidation of alcohol to acetic acid and water

The oxidation of one mole of ethanol yields one mole each of acetic acid and water

C_2H_5OH	$+ O_2 \Rightarrow$	CH_3COOH	+	H_2O
Alcohol		acetic acid		water

5.4 Bacteria of alkaline fermentations

A third group of bacteria are those which bring about alkaline fermentations - the *Bacillus* species. Of note are *Bacillus subtilis, B. licheniformis* and *B. pumilius. Bacillus subtilis* is the dominant species, causing the hydrolysis of protein to amino acids and peptides and releasing ammonia, which increases the alkalinity and makes the substrate unsuitable for the growth of spoilage organisms. Alkaline fermentations are more common with protein rich foods such as soybeans and other legumes, although there are a few examples utilising plant seeds. For example water melon seeds (*Ogiri* in Nigeria) and sesame seeds (*Ogiri-saro* in Sierra Leone) and others where coconut and leaf proteins are the substrates (Indonesian *semayi* and Sudanese *kawal* respectively).

Although the range of products of alkaline fermentation does not match those brought about by acid fermentations, they are important in that they provide protein rich, low cost condiments from leaves, seeds and beans, which contribute to the diet of millions of people in Africa and Asia. Steinkraus presents a comprehensive review of the acid, alkaline and alcoholic fermentations from around the world, which the reader is referred to for further information (Steinkraus, 1996)[87].

5.5 Conditions required for bacterial fermentations

Microorganisms vary in their optimal pH requirements for growth. Most bacteria favour conditions with a near neutral pH (7). The varied pH requirements of different groups of micro-organisms is used to good effect in fermented foods where successions of micro-organisms take over from each other as the pH of the environment changes. Certain bacteria are acid tolerant and will survive at reduced pH levels. Notable acid-tolerant bacteria include *the Lactobacillus* and *Streptococcus* species, which play a role in the fermentation of dairy and vegetable products.

Oxygen requirements vary from species to species. The lactic acid bacteria are described as microaerophilic as they carry out their reactions with very little oxygen. The acetic acid bacteria however, require oxygen to oxidise alcohol to acetic acid. In vinegar production, oxygen has to be made available for the production of acetic acid, whereas with wine it is essential to exclude oxygen to prevent oxidation of the alcohol and spoilage of the wine.

5.5.1 Temperature

Different bacteria can tolerate different temperatures, which provides enormous scope for a range of fermentations. While most bacteria have a temperature optimum of between 20 to 30°C, there are some (the thermophiles) which prefer higher temperatures (50 to 55°C) and those with colder temperature optima (15 to 20°C). Most lactic acid bacteria work best at temperatures of 18 to 22°C. The *Leuconostoc* species which initiate fermentation have an optimum of 18 to 22°C. Temperatures above 22°C, favour the *lactobacillus* species.

5.5.2 Salt concentration

Lactic acid bacteria tolerate high salt concentrations. The salt tolerance gives them an advantage over other less tolerant species and allows the lactic acid fermenters to begin metabolism, which produces acid, which further inhibits the growth of non-desirable organisms. Leuconostoc is noted for its high salt tolerance and for this reason, initiates the majority of lactic acid fermentations.

5.5.3 Water activity

In general, bacteria require a fairly high water activity (0.9 or higher) to survive. There are a few species, which can tolerate water activities lower than this, but usually the yeasts, and fungi will predominate on foods with a lower water activity.

5.5.4 Hydrogen ion concentration (pH)

The optimum pH for most bacteria is near the neutral point (pH 7.0). Certain bacteria are acid tolerant and will survive at reduced pH levels. Notable acid-tolerant bacteria include *the Lactobacillus* and *Streptococcus* species, which play a role in the fermentation of dairy and vegetable products.

5.5.5 Oxygen availability

Some of the fermentative bacteria are anaerobes, while others require oxygen for their metabolic activities. Some, lactobacilli in particular, are microaerophilic. That is they grow in the presence of reduced amounts of atmospheric oxygen. In aerobic fermentations, the amount of oxygen present is one of the limiting factors. It determines the type and amount of biological product obtained, the amount of substrate consumed and the energy released from the reaction. Acetobacter require oxygen for the oxidation of alcohol to acetic acid.

5.5.6 Nutrients

All bacteria require a source of nutrients for metabolism. The fermentative bacteria require carbohydrates – either simple sugars such as glucose and fructose or complex carbohydrates such as starch or cellulose. The energy requirements of microorganisms are very high. Limiting the amount of substrate available can check their growth.

5.6 Principles of lactic acid fermentation

Sauerkraut is one example of an acid fermentation of vegetables. The name sauerkraut literally translates as acid cabbage. The 'sauerkraut process' can be applied to any other suitable type of vegetable product. Because of the importance of this product in the German diet, the process has received substantial research in order to commercialise and standardise production. As a result, the process and the contributing microorganisms are known intimately. Other less well-known fermented fruits and vegetables have received less research attention, therefore little is known of the exact process. It is safe to assume however that the acid fermentation of vegetables is based on this process.

Lactic acid fermentations are carried out under three basic types of condition: – dry salted, brined and non-salted. Salting provides a suitable environment for lactic acid bacteria to grow which impart the acid flavour to the vegetable.

Salt for pickling.

For pickling any variety of common salt is suitable as long as it is pure. Impurities or additives can cause problems. Salt with chemicals to reduce caking should not be used as they make the brine cloudy. Salt with lime impurities can reduce the acidity of the final product and reduce the shelf life of the product. Salt with iron impurities can result in the blackening of the vegetables. Magnesium impurities impart a bitter taste. Carbonates can result in pickles with a soft texture (Lal, Siddappa and Tandon, 1986)[88].

5.6.1 Dry salted fermented vegetables

With dry salting, the vegetable is treated with dry salt. The salt extracts the juice from the vegetable and creates the brine. The vegetable is prepared, washed in potable cold water and drained. For every 100 kg of vegetables 3 kg of salt is needed. The vegetables are placed in a layer of about 2.5cm depth in the fermenting container (a barrel or keg). Salt is sprinkled over the vegetables. Another layer of vegetables is added and more salt added. This is repeated until the container is three quarters full. A cloth is placed above the vegetables and a weight added to compress the vegetables and assist the formation of a brine, which takes about 24 hours. As soon as the brine is formed, fermentation starts and bubbles of carbon dioxide begin to appear. Fermentation takes between one and four weeks depending on the ambient temperature. Fermentation is complete when no more bubbles appear, after which time the pickle can be packaged in a variety of mixtures. These can be vinegar and spices or oil and spices (Lal *et al*, 1986)[89].

5.6.2 The 'sauerkraut' process.

Lactic acid bacteria are the primary group of organisms involved in sauerkraut fermentation. They can be divided into three groups according to their types and end products:

Leuconostoc mesenteroides	an acid and gas producing coccus
Lactobacillus plantarum and *L. Cucumeris*	bacilli that produce acid and a small amount of gas
Lactobacillus pentoaceticus (*L. Brevis*)	acid and gas producing bacilli

In addition to the desirable bacteria there are a range of undesirable microorganisms present on cabbage (and other vegetable material) which can interfere with the sauerkraut process if allowed to multiply unchecked. The quality of the final product depends largely on how well the undesirable organisms are controlled during the fermentation process. Some of the typical spoilage organisms utilise the protein as an energy source, producing unpleasant odours and flavours.

The fermentation process

Shredded cabbage or other suitable vegetables are placed in a jar and salt is added. Mechanical pressure is applied to the cabbage to expel the juice, which contains fermentable sugars and other nutrients suitable for microbial activity. The first microorganisms to start acting are the gas-producing cocci *(L. Mesenteroides).* These microbes produce acids. When the acidity reaches 0.25 to 0.3% (calculated as lactic acid), these bacteria slow down and begin to die off, although their enzymes continue to function. The activity initiated by the *L. mesenteroides* is continued by the lactobacilli *(L. plantarum and L. Cucumeris)* until an acidity level of 1.5 to 2% is attained. The high salt concentration and low temperature inhibit these bacteria to some extent. Finally, *L. pentoaceticus* continues the fermentation, bringing the acidity to 2 to 2.5% thus completing the fermentation.

The end products of a normal kraut fermentation are lactic acid along with smaller amounts of acetic and propionic acids, a mixture of gases of which carbon dioxide is the principal gas, small amounts of alcohol and a mixture of aromatic esters. The acids, in combination with alcohol form esters, which contribute to the characteristic flavour of sauerkraut. The acidity helps to control the growth of spoilage and putrefactive organisms and contributes to the extended shelf life of the product. Changes in the sequence of desirable bacteria, or indeed the presence of undesirable bacteria, alter the taste and quality of the product.

Effects of temperature on sauerkraut process

The optimum temperature for sauerkraut fermentation is around 21°C. A variation of just a few degrees from this temperature alters the activity of the microbial process and affects the quality of the final product. Therefore, temperature control is one of the most important factors in the sauerkraut process. A temperaturc of 18° to 22° C is most desirable for initiating fermentation since this is the optimum temperature range for the growth and metabolism of *L. mesenteroides*. Temperatures above 22°C favour the growth of *Lactobacillus* species.

Effects of salt on the sauerkraut process

Salt plays an important role in initiating the sauerkraut process and affects the quality of the final product. The addition of too much salt may inhibit the desirable bacteria, although it may contribute to the firmness of the kraut. The principle function of salt is to withdraw juice from the cabbage (or other vegetable), thus making a more favourable environment for development of the desired bacteria.

Generally, salt is added to a final concentration of 2.0 to 2.5%. At this concentration, lactobacilli are slightly inhibited, but cocci are not affected. Unfortunately, this concentration of salt has a greater inhibitory effect against the desirable organisms than against those responsible for spoilage. The spoilage organisms can tolerate salt concentrations up to between 5 and 7%, therefore it is the acidic environment created by the lactobacilli that keep the spoilage bacteria at bay, rather than the addition of salt.

In the manufacture of sauerkraut, dry salt is added at the rate if 1 to 1.5 kg per 50kg cabbage (2 to 3%). The use of salt brines is not recommended in sauerkraut making, but is common in vegetables that have a low water content. It is essential to use pure salt since salts with added alkali may neutralise the acid.

Use of starter cultures

In order to produce sauerkraut of consistent quality, starter cultures (similar to those used in the dairy industry) have been recommended. Not only do starter cultures ensure consistency between batches, they speed up the fermentation process as there is no time lag while the relevant microflora colonise the sample. Because the starter cultures used are acidic, they also inhibit the undesirable micro-organisms. It is possible to add starters traditionally used for milk fermentation, such as *Streptococcus lactis*, without adverse effect on final quality. Because these organisms only survive for a short time (long enough to initiate the acidification process) in the kraut medium, they do not disturb the natural sequence of microorganisms. On the other hand, if *Leuconostoc mesenteroides* is added in the early stages, it gives a good flavour to the final product, but alters the sequence of subsequent bacterial growth and results in a product that is incompletely fermented. If gas producing rods (for example *L pentoaceticus)* are added to the sauerkraut, this disturbs the balance between acetic and lactic acids - more acetic acid and less lactic acid are produced than normal - and the fermentation never reaches completion. If lactic acid, non-gas-producing *rods (L. Cucumeris)* are used as a starter, again the kraut is not completely fermented and the resulting product is bitter and more susceptible to spoilage by yeasts.

It is possible to use the juice from a previous kraut fermentation as a starter culture for subsequent fermentations. The efficacy of using old juice depends largely on the types of organisms present in the juice and its acidity. If the starter juice has an acidity of 0.3% or more, it results in a poor quality kraut. This is because the cocci which would normally initiate fermentation are suppressed by the high acidity, leaving the bacilli with sole responsibility for fermentation. If the starter juice has an acidity of 0.25% or less, the kraut produced is normal, but there do not appear to be any beneficial effects of adding this juice. Often, the use of old juice produces a sauerkraut, which has a softer texture than normal.

Spoilage and defects in the sauerkraut process.

The majority of spoilage in sauerkraut is due to aerobic soil microorganisms, which break down the protein and produce undesirable flavour and texture changes. The growth of these aerobes can easily be inhibited by a normal fermentation.

Soft kraut can result from many conditions such as large amounts of air, poor salting procedure and varying temperatures. Whenever the normal sequence of bacterial growth is altered or disturbed, it usually results in a soft product. It is the lactobacilli, which seem to have a greater ability than the cocci to break down cabbage tissues, which are responsible for the softening. High temperatures and a reduced salt content favour the growth of lactobacilli, which are sensitive to higher concentrations of salt. The usual concentration of salt used in sauerkraut production slightly inhibits the lactobacilli, but has no effect on the cocci. If the salt content is too low initially, the lactobacilli grow too rapidly at the beginning and upset the normal sequence of fermentation.

Another problem encountered is the production of dark coloured sauerkraut. This is caused by spoilage organisms during the fermentation process. Several conditions favour the growth of spoilage organisms. For example, an uneven distribution of salt tends to inhibit the desirable organisms while at the same time allowing the undesirable salt tolerant organisms to flourish. An insufficient level of juice to cover the kraut during the fermentation allows undesirable aerobic bacteria and yeasts to grow on the surface of the kraut, causing off flavours and discoloration. If the fermentation temperature is too high, this also encourages the growth of undesirable microflora, which results in a darkened colour.

Pink kraut is a spoilage problem. It is caused by a group of yeasts which produce an intense red pigment in the juice and on the surface of the cabbage. It is caused by an uneven distribution of or an excessive concentration of salt, both of which allow the yeast to multiply. If conditions are optimal for normal fermentation, these spoilage yeasts are suppressed.

5.6.3 Brine salted fermented vegetables

Brine is used for vegetables which inherently contain less moisture. A brine solution is prepared by dissolving salt in water (a 15 to 20% salt solution). Fermentation takes place well in a brine of about 20 salometer. As a general guide, a fresh egg floats in a 10% brine solution (Kordylas, 1990)[90]. Properly brined vegetables will keep well in vinegar for a long time. The duration of brining is important for the overall keeping qualities. The vegetable is immersed in the brine and allowed to ferment. The strong brine solution draws sugar and water out of the vegetable, which decreases the salt concentration. It is crucial that the salt concentration does not fall below 12%, otherwise conditions do not allow for fermentation. To achieve this, extra salt is added periodically to the brine mixture.

Once the vegetables have been brined and the container sealed, there is a rapid development of micro-organisms in the brine. The natural controls which affect the microbial populations of the fermenting vegetables include the concentration of salt and temperature of the brine, the availability of fermentable materials and the numbers and types of micro-organisms present at the start of fermentation. The rapidity of the fermentation is correlated with the concentration of salt in the brine and its temperature.

Most vegetables can be fermented at 12.5° to 20° salometer salt. If so, the microbial sequence of lactic acid bacteria generally follows the classical sauerkraut fermentation described by Pederson (1979)[91]. At higher salt levels of up to about 40° salometer, the sequence is skewed towards the development of a homofermentation, dominated by *Lactobacillus plantarum*. At the highest concentrations of salt (about 60° salometer) the lactic fermentation ceases to function and if any acid is detected during brine storage it is acetic acid, presumably produced by acid-forming yeasts which are still active at this concentration of salt (Vaughn, 1985)[92].

Brine salted fermentation of vegetables (Pickles)

Pickled cucumbers are another fermented product that has been studied in detail and the process is known. The fermentation process is very similar to the sauerkraut process, only brine is used instead of dry salt. The washed cucumbers are placed in large tanks and salt brine (15 to 20%) is added. The cucumbers are submerged in the brine, ensuring that none float on the surface - this is essential to prevent spoilage. The strong brine draws the sugar and water out of the cucumbers, which simultaneously reduces the salinity of the solution. In order to maintain a salt solution so that fermentation can take place, more salt has to be added to the brine solution. If the concentration of salt falls below 12%, it will result in spoilage of the pickles through putrefaction and softening.

A few days after the cucumbers have been placed in the brine, the fermentation process begins. The process generates heat which causes the brine to boil rapidly. Acids are also produced as a result of the fermentation.

During fermentation, visible changes take place which are important in judging the progress of the process. The colour of the cucumber surface changes from bright green to a dark olive green as acids interact with the chlorophyll. The interior of the cucumber changes from white to a waxy translucent shade as air is forced out of the cells. The specific gravity of the cucumbers also increases as a result of the gradual absorption of salt and they begin to sink in the brine rather than floating on the surface.

Microbes involved in the fermentation process

As with the sauerkraut process, the gram positive coccus - *Leuconostoc mesenteroides* predominates in the first stages of pickle fermentation. This species is more resistant to temperature changes and tolerates higher salt concentration than the subsequent species. As fermentation proceeds and the acidity increases, lactobacilli start to take over from the cocci. The active stage of fermentation continues for between 10 to 30 days, depending upon the temperature of the fermentation. The optimum temperature for *L. Cucumeris* is 29 to 32°C. During the fermentative period, the acidity increases to about 2% and the strong acid producing types of bacteria reach their maximum growth. If sugar or acetic acid is added to the fermenting mixture during this time it increases the production of acid.

Problems in pickles

The production of excessive amounts of acid during the fermentation, results in shrivelling of the pickles, possibly due to over-activity of the *L. mesenteroides* species. If the brine is stirred, it may introduce air, which makes conditions more favourable for the growth of spoilage bacteria. In general, if the pickles are well covered with brine, the salt concentration is maintained and the temperature is at an optimum, it should be quite simple to produce good quality pickles.

5.6.4 Non salted, lactic acid fermented vegetables

Some vegetables are fermented by lactic acid bacteria, without the prior addition of salt or brine. Examples of non-salted products include *gundruk* (consumed in Nepal), *sinki* and other wilted fermented leaves. The detoxification of cassava through fermentation includes an acid fermentation, during which time the cyanogenic glycosides are hydrolysed to liberate the toxic cyanide gas.

The fermentation process relies on the rapid colonisation of the food by lactic acid producing bacteria, which lower the pH and make the environment unsuitable for the growth of spoilage organisms. Oxygen is also excluded as the *Lactobacilli* favour an anaerobic atmosphere. Restriction of oxygen ensures that yeasts do not grow.

For the production of *sinki*, fresh radish roots are harvested, washed and wilted by sun drying for one to two days. They are then shredded, re-washed and packed tightly into an earthenware or glass jar, which is sealed and left to ferment. The optimum fermentation time is twelve days at 30°C. *Sinki* fermentation is initiated by *L. fermentum* and *L. brevis*, followed by *L. plantarum*. During fermentation the pH drops from 6.7 to 3.3. After fermentation, the radish substrate is sun-dried to a moisture level of about 21%. For consumption, *sinki* is rinsed in water for two minutes, squeezed to remove the excess water and fried with salt, tomato, onion and green chilli. The fried mixture is then boiled in rice water and served hot as soup along with the main meal (Steinkraus, 1996)[93].

Pit fermentations

South Pacific pit fermentations are an ancient method of preserving starchy vegetables without the addition of salt. The raw materials undergo an acid fermentation within the pit, to produce a paste with good keeping qualities. Pit fermentations are also used in other parts of the world – for example in Ethiopia, where the false banana (*Ensete ventricosum*) is fermented in a pit to produce a pulp known as *kocho*. Foods preserved in pits can last for years without deterioration; therefore pits provide a good, reliable cheap means of storage.

Root crops and bananas are peeled before being placed in the pit, while breadfruit are scraped and pierced. Food is left to ferment for three to six weeks, after which time it becomes soft has a strong odour and a paste-like consistency. During fermentation, carbon dioxide builds up in the pit, creating an anaerobic atmosphere. As a result of bacterial activity, the temperature rises much higher than the ambient temperature. The pH of the fruit within the pit decreases from 6.7 to 3.7 within about four weeks. Inoculation of the fruit in the pit with lactic acid bacteria greatly speeds up the process. The fermented paste can be left in the pit and removed as required. Usually, it is removed and replaced with a second batch of fresh food to ferment. The fermented food is washed and fibrous material removed. It is then dried in the sun for several hours to remove the volatile odours, and pounded into a paste. Grated coconut or coconut cream and sugar may be added and the mixture is wrapped in banana leaves and either baked or boiled (Steinkraus, 1996)[94].

5.7 Principles of Acetic Acid Fermentation

The main desirable fermentation carried out by acetic acid bacteria is the production of vinegar. Vinegar, literally translated as sour wine, is one of the oldest products of fermentation used by man. It can be made from almost any fermentable carbohydrate source, for example fruits, vegetables, syrups and wine. Whatever the raw material used, the fermentation process follows a definite sequence.

The basic requirement for vinegar production is a raw material that will undergo an alcoholic fermentation. Apples, pears, grapes, honey, syrups, cereals, hydrolysed starches, beer and wine are all ideal substrates for the production of vinegar. The best raw materials are cider and wine, which are widely used in Europe and the United States. To produce a high quality product it is essential that the raw material is mature, clean and in good condition.

5.7.1 Microbes involved in the vinegar process.

The production of vinegar depends on a mixed fermentation, which involves both yeasts and bacteria. The fermentation is usually initiated by yeasts, which break down glucose into ethyl alcohol with the liberation of carbon dioxide gas. Following on from the yeasts, *acetobacter* oxidise the alcohol to acetic acid and water.

Yeast reaction

$C_6H_{12}O_6$	$\Rightarrow$	$2C_2H_5OH$	+	$2CO_2$
Glucose	yeast	ethyl alcohol	+	carbon dioxide

Bacterial reaction

C_2H_5OH	+	O_2	$\Rightarrow$	CH_3COOH	+	H_2O
Alcohol				acetic acid		water

The yeasts and bacteria exist together in a form known as commensalism. The acetobacter are dependent upon the yeasts to produce an easily oxidisable substance (ethyl alcohol). It is not possible to produce vinegar by the action of one type of microorganism alone.

For a good fermentation, it is essential to have an alcohol concentration of 10 to 13%. If the alcohol content is much higher, the alcohol is incompletely oxidised to acetic acid. If it is lower than 13%, there is a loss of vinegar because the esters and acetic acid are oxidised. In addition to acetic acid, other organic acids are formed during the fermentation which become esterified and contribute to the characteristic odour, flavour and colour of the vinegar.

Acetaldehyde is an intermediate product in the transformation of the reducing sugar in fruit juice to acetic acid or vinegar. Oxygen is required for the conversion of acetaldehyde to acetic acid.

In general, the yield of acetic acid from glucose is approximately 60%. That is three parts of glucose yield two parts acetic acid.

5.7.2 Microorganisms involved in the fermentation of vinegar.

The organisms involved in vinegar production usually grow at the top of the substrate, forming a jelly like mass. This mass is known as 'mother of vinegar'. The mother is composed of both *acetobacter* and yeasts, which work together. The principal bacteria are *Acetobacter acetic A. Xylinum* and *A. Ascendens*. The main yeasts are *Saccharomyces ellipsoideus* and S *cerevisiae.* It is important to maintain an acidic environment to suppress the growth of undesirable organisms and to encourage the presence of desirable acetic acid producing bacteria. It is common practice to add 10 to 25% by volume of strong vinegar to the alcoholic substrate in order to attain a desirable fermentation.

The alcoholic fermentation of sugars should be completed before the solution is acidified because any remaining sugar will not be converted to alcohol after the acetic acid is added. Incomplete fermentation of the juice results in a "weak" product. The acetic acid strength of good vinegar should be approximately 6%.

5.7.3 Fermentation methods

Small scale production

Vinegar can be made at home at the small scale by introducing oxygen into barrels of wine or cider and allowing fermentation to occur spontaneously. This process is not very rigorously controlled and often results in a poor quality product.

The Orleans process

The Orleans process is one of the oldest and well-known methods for the production of vinegar. It is a slow, continuous process, which originated in France. High-grade vinegar is used as a starter culture, to which wine is added at weekly intervals. The vinegar is fermented in large (200 litre) capacity barrels. Approximately 65 to 70 litres of high-grade vinegar are added to the barrel along with 15 litres of wine. After one week, a further 10 to 15 litres of wine are added and this is repeated at weekly intervals. After about four weeks, vinegar can be withdrawn from the barrel (10 to 15 litres per week) as more wine is added to replace the vinegar.

One of the problems encountered with this method is that of how to add more liquid to the barrel without disturbing the floating bacterial mat. This can be overcome by using a glass tube, which reaches, to the bottom of the barrel. Additional liquid is poured in through the tube and therefore does not disturb the bacteria. Wood shavings are sometimes added to the fermenting barrel to help support the bacterial mat.

Quick vinegar method

Because the Orleans process is slow, other methods have been adapted to try and speed up the process. The German method is one such method. It uses a generator, which is an upright tank filled with beechwood shavings and fitted with devices, which allow the alcoholic solution to trickle down through the shavings in which the acetic acid bacteria are living. The tank is not allowed to fill, as that would exclude oxygen, which is necessary for the fermentation. Near the bottom of the generator are holes which allow air to be drawn in. the air rises through the generator and is used by the acetic acid bacteria to oxidise the alcohol. This oxidisation also releases considerable amounts of heat, which must be controlled to avoid causing damage to the bacteria.

Problems in vinegar production

Many of the problems of vinegar production are concerned with the presence of nematodes, mites, flies and other insects. These pests can be controlled by adherence to good hygiene and pasteurisation of the vinegar. Problems associated with the fermentative process include the presence of a whitish film on the surface of the vinegar. This is sometimes called *Mycoderma vini* and is composed of yeast-like organisms, which grow aerobically and oxidise the carbon containing compounds to carbon dioxide and water. They also alter the flavour and alcohol content of the vinegar. This problem can however be controlled by adding one part vinegar to three parts of the alcoholic solution or by storing the alcoholic liquid in filled closed containers.

CHAPTER 6

TYPICAL PRODUCTS OF BACTERIAL FERMENTATIONS

Many products are produced by bacterial fermentations. These include the fruit and vegetable pickles produced by lactic acid fermentation and the products of alkaline bacterial fermentations. Lactic acid bacteria pickling is still carried out at the domestic scale. However industrial scale processes have been developed for the production of most types of pickles. Pickles can be made by storing prepared vegetables in a weak brine solution, by dry salting or allowing the vegetables to ferment without salt.

6.1 Dry salted pickles

Dry salting is used for pickling many vegetables and fruits including limes, lemons and cucumbers. For dry salt pickling any variety of common salt is suitable as long as it is pure. Impurities or additives can cause problems:

- Chemicals to reduce caking should not be used as they make the brine cloudy.
- Lime impurities can reduce the acidity of the final product and reduce the shelf life of the product.
- Iron impurities can result in the blackening of the vegetables.
- Magnesium impurities impart a bitter taste.
- Carbonates can result in pickles with a soft texture (Lal *et al*, 1986)[95].

6.1.1 Dry salted lime pickle.

Location of production

Dry salted lime pickles are produced in Asia and Africa. They are particularly popular in India, Pakistan and North Africa.

Product Description

With dry salting, the limes are treated with dry salt. The salt extracts and juice from the vegetable and create the brine. The final product is a sour lime pickle. Spices are added depending on local preference. In India and Pakistan, the pickle is usually very spicy and hot due to the addition of chilli. It is usually eaten as a condiment.

Preparation of raw materials

The limes need to be selected and prepared. Only fully ripe limes without bruising or damage should be used. All limes need to be washed in potable cold water, drained, and then cut into quarters. Spices should be of good quality and free of mould.

Processing

Limes are placed in a layer, approximately 2.5 cm deep, into the fermenting container (a barrel or keg). One kilogram of salt is added for every four kilograms of limes. The salt is sprinkled over the vegetables. Another layer of vegetables is added and more salt added. This is repeated until the container is three quarters full. A cloth is placed above the vegetables and a weight added to compress the vegetables and assist in the formation of brine. The formation of brine takes about 24 hours. The container is then placed in the sun for a week.

As soon as the brine is formed, fermentation starts. As fermentation starts bubbles of carbon dioxide appear. Fermentation takes between one and four weeks depending on the ambient temperature. Fermentation is complete when no more bubbles appear. (Lal *et al*, 1986)[96], (Kordylas, 1990)[97].

Flow diagram

Selection	Only ripe limes should be selected
⇓	
Wash	In clean water
⇓	
Cut	Cut into four pieces or slice the skin
⇓	
Mix with salt	1kg salt for 4kg of limes
⇓	
Ferment	Leave the container in the sun for a week to ferment
⇓	
Package	

Packaging and storage

The vegetables can be removed from the brine and packaged in a variety of mixtures, which may consist of vinegar and spices or oil and spices. Lime pickle can be packed in small polythene bags and sealed or in clean jars and capped. Lime pickles keep well if stored in a cool place. Due to the high acid level of the final product, the risk of food poisoning is low (Fellows, 1997)[98].

6.1.2 Pickled cucumbers

Location of production

Pickled cucumbers are made in Africa, Asia and Latin America.

Product description

Cucumbers undergo typical lactic acid fermentation and change from a pale product to a darker green and more transparent product. *Khalpi* is a cucumber pickle popular during the summer months in Nepal (Karki, 1986)[99].

Raw material preparation

Fully ripe cucumbers without bruising or damage are washed in potable cold water and drained. The cucumbers can be pickled whole or sliced. With *khalpi* the cucumbers are washed, sliced and cut into 5-8 cm pieces.

Processing

1 kg of salt is added to every 20-kg of small cucumbers and 15 kg of large cucumbers. The brine should be formed within 24 hours by osmosis. If the brine formed by osmosis does not cover the cucumbers 40° Salometer brine is added to the desired level. A day or two after the tank is filled and closed the brine should be stirred in order to help equalise the concentration of salt throughout the mass (Vaughn, 1985)[100].

As soon as the brine is formed, fermentation starts and bubbles of carbon dioxide appear. Fermentation takes between one and four weeks depending on the ambient temperature. Fermentation is complete when no more bubbles appear.

During fermentation the brine becomes cloudy for the first few days due to the growth of bacteria. Later if the brine is not covered, a filmy yeast growth will often occur on the surface (Pedersen, 1979)[101]

Flow diagram

Selection	Only ripe cucumbers should be selected
⇓	
Wash	In clean water
⇓	
Mix with salt	1kg salt for 15-20kg of cucumbers
⇓	
Ferment	For between one and four weeks
⇓	
Package	

Packaging and storage

Cucumber pickle is usually stored in clean capped jars. They keep well if stored in a cool place. Due to the high acid level of the final product, the risk of food poisoning is low. With *khalpi* in Nepal, oil is added.

6.1.3 *Pak-Gard-Dong* (Pickled leafy vegetable)

Pak-Gard-Dong is a fermented mustard leaf (*Brassica juncea*) product made in Thailand. The mustard leaves are washed, wilted in the sun, mixed with salt, packed into containers for 12 hours. The water is then drained and a 3% sugar solution added. They are again allowed to ferment for three to five days at room temperature. Microorganisms associated with the fermentation include *Lactobacillus brevis, Pediococcus cerevisiae* and *Lactobacillus plantarum* (Boon-Long, 1992)[102].

A similar product (*Hum choy*) is made in the South of China. This is produced by fermenting a local leafy vegetable. The leaves are washed and drained. They are then covered in salt and hung on racks to dry in the sun. The wilted leaves are placed in earthenware pots and covered with rice water, obtained after washing rice grains. The pots are sealed and the leaves allowed to ferment for four days. The product can be stored for up to two months if the seal is not broken (Steinkraus, 1996)[103].

6.1.4 *Tempoyak (*pickled durian)

Tempoyak is the fermented pulp of a durian fruit (*Durio zibethinus*) from Malaysia. It has the distinctive durian smell and a creamy yellow colour. It is made by mixing durian pulp with salt and placing in a sealed container. Fermentation takes about seven days.

6.1.5 Pickled beetroots

In Russia beetroot is pickled by cleaning, slicing and placing in a container with salt. Due to the high sucrose level, dextrans are produced giving the product a slimy texture (Pedersen, 1979)[104].

6.1.6 Lamoun Makbous (pickled lemons)

Pickled lemons are popular in Asia. In west Asia and North Africa they are known as *lamoun makbous* and *msir*. Lemons are washed in clean water, sliced and covered in salt. After at least 24 hours, they are drained and mixed with oil and spices.

6.2 Brined fruit and vegetable pickles

For brine pickling any variety of common salt is suitable as long as it is pure. Impurities or additives can cause problems:

- Chemicals to reduce caking should not be used as they make the brine cloudy.
- Lime impurities can reduce the acidity of the final product and reduce the shelf life of the product.
- Iron impurities can result in the blackening of the vegetables.
- Magnesium impurities impart a bitter taste.
- Carbonates can result in pickles with a soft texture (Lal *et al*, 1986)[105].

6.2.1 Green Mango Pickle.

Location of production

Mango pickle is a very popular pickle in many Asian, African and Latin American countries. It is a major product of India, Pakistan and Bangladesh and it is estimated that the annual production of mango pickle in South Africa is over 10,000 tons (Redelinghuys and Van der Riet, 1978)[106].

Product description

Green mango pickle is a hot, spicy pickle with a sour taste. It is eaten as a condiment. Preservation is caused by a combination of salt, increased acidity and to a small extent the spices. It is known as *burong mangga* and *dalok* in the Philippines.

Preparation of the raw material

The fresh, fully mature, firm but unripe mangoes must be carefully selected to ensure a good quality product. The best pickles are obtained from fruit at early maturity when the fruit has reached almost maximum size. Riper fruit results in pickles with a fruity odour and lacking the characteristic and predominant green mango flavour.

The green mangoes need to be inspected and any damaged fruit rejected. The fruit is washed in clean water and drained.

After draining, the fruit is cut. Sharp knives with preferably stainless blades should be used. Iron or copper equipment should be avoided. A single stroke should be used during the cutting process to ensure minimum damage and avoiding mushiness in the final product.

Processing

The sliced mangoes are soaked in brine solution. Sodium metabisulphite (1000 ppm) and 1% calcium chloride are added. The containers are stored until the mangoes are pickled. The brine is then drained off and spices are mixed with the mango slices (Redelinghuys and Van der Riet, 1978)[107].

Flow diagram

Fruit	The mangoes need to be unripe
⇓	
Sort	Remove damaged fruit
⇓	
Wash	With clean water.
⇓	
Drain	
⇓	
Cut	Stainless steel knives should be used
⇓	
Soaked in brine	20% salt solution
⇓	
Drain	
⇓	
Add spices	Spices to taste
⇓	
Pack	Pack in containers and add oil

Packaging and storage

The mixture is then packed and oil added onto the surface of the mixture. The mangoes should be firmly pressed down in the container. Good quality vegetable oil such as sunflower oil should be used and finely ground chilli powder can be added to the oil for flavour and colour. Mango pickle can be packed in small polythene bags and sealed or in clean jars and capped. Mango pickle keeps well if stored in a cool place. If it is processed well, it can be kept for several months. Due to the high acid level of the final product, the risk of food poisoning is low (Fellows, 1997)[108].

6.2.2 Lime pickle (brined).

Location of production

Lime pickles are produced in Asia, Latin America and Africa. They are particularly popular in India, Pakistan and North Africa.

Product description

Lime pickle is made from salted pieces of lime packed in a salty, spicy liquor, like a semi-solid gravy. It is brownish red and the lime peels are yellow or pale green with a sour and salty taste. It is eaten as a condiment with curries or other main meals. If processed well, the product can be kept for several months.

Preparation of raw materials

The limes need to be selected and prepared. Only fully ripe limes without bruising or damage should be used. All the limes need to be washed in potable cold water and drained. The limes are dipped in hot water (60-65^{0}C) for about five minutes. They are then cut into pieces in order to expose the interior and allow salt to be absorbed more quickly.

All spices should be of good quality and free of mould.

Processing

The prepared limes are covered with a brine solution. This causes water to be drawn out of the pieces by osmosis. It is important to ensure that the surface is covered with juice, and leave for 24 hours. If necessary, the fruits should be pressed down to hold them below the liquid. Once the limes have been placed in the brine, there is a rapid development of microorganisms and fermentation begins.

After fermentation the limes are dried in the sun until the skin becomes brown.

Flow diagram

Sort — Select ripe (but not overripe) healthy lime fruits.
⇓
Wash
⇓
Heat — Dip in hot water (60-65^{0}C) for about five minutes.
⇓
Cut — Cut into four pieces or alternatively cut into smaller, uniform-sized pieces.
⇓
Brine — Ensure that the surface is covered with juice.
⇓
Dry — Dry in the sun for 2-3 days.
⇓
Mix spices — To local preference
⇓
Pack
⇓
Store — In a cool place, away from sunlight.

Packaging and storage

The limes are mixed with spices and oils according to local taste and tradition. Lime pickle can be packed in small polythene bags and sealed or in clean jars and capped. Lime pickle keeps well if stored in a cool place. Due to the high acid level of the final product, the risk of food poisoning is low.

6.2.3 *Kimchi* (pickled cabbage)

Location of production

Kimchi is probably the most important processed food product in Korea. It is an essential dish, eaten at most mealtimes. Production is estimated at over one million tons, mainly at household level. Daily consumption is estimated at 150 to 250 grams per person.

Product description

Kimchi is a general name for a range of closely related fermented products. It is similar to Sauerkraut in Europe and the United States. There are numerous variations of *kimchi* depending on the production technique. The main pickled cabbage *kimchis* are *tongbaechu-kimchi tongkimchi* and *bossam-kimchi.* This section refers to *kimchi* produced from cabbage; the following section deals with pickled radish products.

Preparation of raw materials

Appropriate cultivars of Chinese cabbage, with light-green coloured soft leaves and compact structures with no defects, are required for production of *kimchi*. After removing outer leaves and roots from the cabbage, it is cut into small pieces.

Processing

The prepared cabbage is placed in a salt solution (8-15%) for two to seven hours in order increase the salt content of the cabbage to between 2.0-4.0% (w/w). It is then rinsed several times with fresh water and drained to remove extra water by centrifugation or by allowing to stand.

Kimchi fermentation is carried out by various microorganisms present in the raw materials and ingredients used in the preparation of *kimchi*. Among the two hundred bacteria isolated form kimchi, the important microorganisms in *kimchi* fermentation are known to be *Lactobacillus plantarum, L. Brevis, Streptococcus faecalis, Leuconostoc mesenteroides* and *Pediococcus pentosaceus.*

After fermentation, the product can be left to mature for several weeks if refrigeration is available. If stored under warm conditions, the *kimchi* deteriorates rapidly.

Flow Diagram

Korean cabbage	
⇓	
Cut	Cut into four parts
⇓	
Add brine	5% for 12 hours or 15% for 7hours
⇓	
Rinse	Drain water
⇓	
Mix	With salt, spices and flavourings
⇓	
Ferment	

6.2.4 Green olives

Olives are a brine fermented product which undergo an essential pre-treatment with lye to remove substances which are toxic to bacteria and would, if left in the olives, prevent fermentation.

Green olives are placed in a 2% sodium hydroxide (lye) solution at 21^0 to 24^0C until the lye penetrates the flesh. Cold water is added to the solution, which dilutes the mixture until the lye is completely removed. The lye treatment is necessary to remove a bitter glucoside compound (oleuropein) from the outer tissues of the olive. Oleuropein is highly toxic to bacteria and therefore needs to be removed in order for a fermentation to take place. After the glucoside has been removed, the olives are placed in barrels with a 1 to 10% brine solution and allowed to undergo a spontaneous fermentation. The optimum fermentation temperature is 24^0C. The fermentation period usually takes between two and three months. Once fermentation is complete, the olives are packed in airtight jars and sterilised which produces a good quality product with a long storage life.

6.2.5 Black olives

When ripe olives are used, they are placed in a 5 to 7% brine solution to soften the outer tissue before lye is added. This pre-treatment allows the lye to penetrate the fruit more easily. In ripe fruit, the glucoside is situated more deeply within the tissue. After soaking in brine, the lye solution (0.7 to 2.0%) is added and, after soaking, the olives are washed clean. They are then packed in barrels with a 2 to 5% brine solution and allowed to undergo fermentation for two to six weeks, depending on the external temperature. Unlike immature olives, the ripe one are exposed to air during fermentation. This is to oxidise the polyphenols in the tissue to a black colour, which is dependent upon oxygen and the small amounts of sodium hydroxide, which are left on the olives. The finished product is packed in a 3 to 5% brine solution and then sterilised. *Bacillus subtilis* may produce pectolytic enzymes, which result in a soft product. If the wash water used for removing the lye is below $60^{0}C$, many species of undesirable bacteria can survive which result in a low quality product.

6.2.6 Jackfruit pickle

Young green jackfruit is pickled in India and Sri Lanka. Young green jackfruit are peeled and cut into 1.2 to 1.8-cm thick slices. The slices are placed in a container and covered in an 8% common salt solution. They are weighed down to keep them submerged in the brine. The brine solution is increased by 2% each day until it reaches 15%. The slices are then left for 8-10 days in the brine. Vinegar and spices are added prior to packaging (Lal, Siddappa, and Tandon, 1986)[109].

6.2.7 Pickled Radish

A number of pickled radish products are produced in Korea. These include **kaktugi, tongchimi, chonggak**-*kimchi, seokbakji, yolmu-kimchi dan moogi kach doo ki gactuki and mootsanji*.

6.2.8 Pickled cucumber

A variety of brine pickled cucumber products are made around the world. *Oi sobagi* and *oiji* are made in Korea. In Egypt cucumbers are pickled by soaking in brine to produce *torshi khiar.*

6.2.9 Pickled leafy vegetables

There are many other brine pickled leafy vegetables around the world. For instance:

- *Pak-sian-dong*: This is a popular pickled leafy vegetable *Pak Sian* (*Gynadropsis pentaphylla*) in Thailand. The fresh vegetable is cleaned and wilted in the sun for one to two hours. It is then placed in brine and fermented for two to three days at room temperature
- *Sayur asin* fermented wilted mustard cabbage (*Brassica juncea*) from Indonesia. It is also known as *kiam chai* in Thailand and *kiam chaye* in Malaysia.

6.2.10 Other pickled vegetables and fruits

- *Naw-mai-dong*: pickled bamboo shoots (*Bamhusa glaucescens*) from Thailand

- *Hom-dong*: pickled red onions from Thailand
- Jeruk: pickled vegetables including ginger and papaya from Malaysia
- Pickled carrots and turnips are produced in Asia and Africa. They are known as *hua-chai po* in Thailand and *tai tan tsoi* in China.
- *Nukamiso-zuke*: vegetables fermented in rice bran, salt and water in Japan (Campbell Platt, 1987)[110]
- Bananas are pickled in the West Indies
- Fermented sweet peppers (*torshi felfel*) are produced in west Asia and Africa.
- Cauliflower stalks are fermented to produce *achar tandal* in India.
- Aubergines (*torshi betingen*) are pickled in west Asia.

6.3 Non salted lactic acid bacteria products

6.3.1 *Gundruk* (pickled leafy vegetable)

Location of production

Gundruk is particularly popular in Nepal. The annual production of *gundruk* in Nepal is estimated at 2,000 tons and most of the production is carried out at the household level.

Product description

Gundruk is obtained from the fermentation of leafy vegetables in Nepal. It is served as a side dish with the main meal and is also used as an appetiser. *Gundruk* is an important source of minerals particularly during the off-season when the diet consists of mostly starchy tubers and maize which tend to be low in minerals (Karki, 1986)[111].

Preparation of raw materials

In the months of October and November, during the harvest of the first broad mustard, radish and cauliflower leaves, large quantities of leaves accumulate - much more than can be consumed fresh (Aidoo, 1986)[112]. These leaves are allowed to wilt for one or two days and then shredded with a knife or sickle.

Processing

Shredded leaves are tightly packed in an earthenware pot and warm water (at about 30°C) is added to cover all the leaves. The pot is then kept in a warm place. After five to seven days, a mild acidic taste indicates the end of fermentation and the *gundruk* is removed and sun-dried. This process is similar to sauerkraut production except that no salt is added to the shredded leaves prior to *gundruk* fermentation. The ambient temperature at the time of fermentation is about 18°C (Jones, 1994)[113], (Karki, 1986)[114].

Pediococcus and *Lactobacillus* species are the predominant microorganisms during *gundruk* fermentation. The fermentation is initiated by *L. cellobiosus* and *L. plantarum*, and other homolactics make a vigorous growth from the third day onwards. *Pediococcus pentosaceus* increases in number on the fifth day and thereafter declines (Karki *et al,*

1983)[115]. During fermentation, the pH drops slowly to a final value of 4.0 and the amount of acid (as lactic) increases to about 1% on the sixth day.

It has been found that a disadvantage with the traditional process of *gundruk* fermentation is the loss of 90% of the carotenoids, probably during sun drying. Improved methods of drying might reduce the vitamin loss (Aidoo, 1992)[116].

Flow diagram

Leafy vegetables	
⇓	
Wilt	One to two days
⇓	
Shred	
⇓	
Placed in earthen pot	The leaves need to tightly packed
⇓	
Cover the leaves	Cover the leaves with warm water and straw
⇓	
Ferment	The pot is kept warm in the sun and by a fire by night.
⇓	
Add warm water	To keep the pot warm
⇓	
Dried	Product dried on mats in the sun

6.3.2 *Kocho* (pickled false banana)

Location of production

Ethiopia

Product description

False banana (*Ensete ventricosum*) is fermented in a pit to produce a pulp known as *kocho*. Foods preserved in pits can last for years without deterioration. Pits therefore provide a good, reliable cheap means of storage.

Processing

The type of soil and its drainage are important in the selection of a pit site. Pits are often lined with stones to prevent the soil from the sidewalls falling into the bottom. A family pit may be 0.6 to 1.5 metres deep and 1.2 to 2 metres wide with a capacity of about fifty breadfruits. A community pit is usually much larger, with the capacity to hold up to 1000 breadfruit. A family pit requires at least 1000 green banana leaves and four sacks of dried banana leaves for lining the walls and top. It is essential that proper attention be given to hygiene of the pit and the fruit to be stored in it.

The central stems are removed from fresh banana leaves and they are wilted in the sun until they become soft and pliable. The pit is lined with dry leaves, then green leave are folded and arranged, overlapping each other, around the sides of the pit and extending over the top. At lest two or three layers of banana leaves are used to seal the pit and prevent contamination by the soil. Washed, peeled food is placed in the pit, green banana leaves are folded over the top of the food and heavy stones are placed on top to weigh down the leaves.

Root crops and bananas are peeled before placing in the pit, breadfruit are scraped and pierced. Food is left to ferment for three to six weeks, after which time it becomes soft has a strong odour and a paste-like consistency. The fermented paste can be left in the pit and removed as required. Usually, it is removed and replaced with a second batch of fresh food to ferment. The fermented food is washed and fibrous material removed. It is then dried in the sun for several hours to remove the volatile odours. It is then pounded into a paste. Grated coconut or coconut cream and sugar added and the mixture is wrapped in banana leaves and either baked or boiled (Steinkraus, 1996)[117].

During fermentation, carbon dioxide builds up in the pit, creating an anaerobic atmosphere. As a result of bacterial activity, the temperature rises much higher than the ambient temperature. The pH of the fruit within the pit decreases from 6.7 to 3.7 within about four weeks. Inoculation of the fruit in the pit with lactic acid bacteria greatly speeds up the process.

In the South Pacific pit fermentations are an ancient method of preserving starchy vegetables such as banana, plantain, breadfruit, cassava, taro, sweet potato, arrowroot and yams. The products undergo an acid fermentation, to produce a paste with good keeping qualities. It is usually pounded with a little sugar, coconut cream or fresh coconut and boiled or baked to make a type of pudding.

*6.3.3 Sinki (*pickled radish*)*

Sinki is a sour pickle prepared from radish taproots. It is consumed traditionally in India, Nepal and parts of Bhutan, where it is used as a base for soup or eaten as a pickle. It is one of the most popular pickles in Nepal. Fresh radish roots are harvested, washed and wilted by sun drying for one to two days. They are then shredded, re-washed and packed tightly into an earthenware or glass jar, which is sealed and left to ferment. The optimum fermentation time is twelve days at 30°C. *Sinki* fermentation is initiated by *L. fermentum* and *L. brevis*, followed by *L. plantarum*. During fermentation the pH drops from 6.7 to 3.3. After fermentation, the radish substrate is sun-dried to a moisture level of about 21% (Steinkraus, 1996)[118]. There is a second processing method involving fermentation in a clay lined pit for two to three months (Karki, 1986)[119]. For consumption, *sinki* is rinsed in water for two minutes, squeezed to remove the excess water and fried with salt, tomato, onion and green chilli. The fried mixture is then boiled in rice water and served hot as soup along with the main meal (Steinkraus, 1996)[120].

6.3.4 *Sunki*

Sunki is a non-salted and fermented vegetable product prepared from the leaves of "Otaki-turnip" in Kiso district, Nagano prefecture, Japan. *Sunki* is eaten with rice and in *miso* soup. The Otaki-turnip is boiled, inoculated with "*Zumi*" (a wild small apple) dried *Sunki* from the previous year and allowed to ferment for one to two months. *Sunki* is produced under low temperature (in winter season). Microorganisms involved include *Lactobacillus plantarum, L. Brevis, Bacillus coagulans and Pediococcus pentosaceus* (Makayama, 1957)[121].

6.3.5 *Kanji*

In Northern India and Pakistan carrots, especially a variety that is deep purple in colour, are fermented to make a traditional ready to serve drink known as *kanji.* Kanji is very popular and considered to have cooling and soothing properties and to be of high nutritional value. After thorough washing the carrots are finely grated. Each kilogram of grated carrot is mixed with 7 litres of water, 200g of salt, 40g of crushed mustard seed and 8g of hot chilli powder. The mixture is then placed in a glazed earthenware vessel, which is almost entirely sealed, leaving only a tiny hole for gases released during fermentation to escape. The mixture is then allowed to ferment for seven to ten days. The type of fermentation that takes place is known as a lactic fermentation, which must be carried out in the absence of air. Lactic acid bacteria produce lactic acid which reduces the pH (i.e. increases the acidity) to a level that prevents the growth of food poisoning organisms. The final product is slightly acidic in taste and has an attractive purple-red colour. After fermentation the drink is strained through fine muslin and has to be consumed within 3 or 4 days after which it goes bad. Each kg of grated carrot yields just over 7 litres of *kanji (*Berry, 1998)[122], (Shah, 1986)[123].

6.3.6 Fermented tea leaves

In South East Asia, tealeaves (*Camellia sinensis*) are fermented to make a sour-tasting snack. In Meaner the product is called *leper-so*, in Thailand it is known as *miang*.

6.4 Alkaline fermentation products

6.4.1 Kawal

Location of production

Sudan

Product description

Kawal is a strong smelling Sudanese, protein-rich food prepared by fermenting the leaves of a wild African legume, *Cassia obtusifolia* and is usually cooked in stews and soups. It is used as a meat replacer or a meat extender. Its protein is of high quality, rich in sulphur amino acids which are usually obtained from either fish or meat (Dirar, 1992)[124].

Raw material preparation

The *Sickle Pod* plant (*Cassia obtusifolia*) is a wild legume that grows in Sudan. The leaves are collected late in the rainy season when the plant is fully-grown. All the stems, pods and flowers are removed. The leaves are not washed, since it is thought that natural microorganisms on the leaves are important for the correct fermentation.

Process and principles of preservation

The leaves of the leguminous plant are pounded into paste without releasing the juice. The paste is placed in an earthenware jar and covered with sorghum leaves. The whole jar is sealed with mud and buried in the ground up to the neck in a cool place. Every three days the contents are mixed by hand.

The fermentation takes about fourteen days. The fermentation is extremely complex. The main microorganisms are *Bacillus subtilis* and *Propionibacterium spp.* Lactic acid bacteria including *Lactobacillus plantarum* ; yeasts including *Candida krusei* and *Saccharomyces spp* and moulds including *Rhizopus spp* are also involved.

After about fourteen days, the strongly smelling black fermented paste is made into small balls and sun-dried for five days. (Harper and Collins, 1992)[125], (Aidoo, 1986),[126]

Flow diagram

Select leaves	Remove stems and flowers
⇓	
Grind leaves	Pound the leaves into a paste in a mortar and pestle
⇓	
Place in jar	
⇓	
Cover jar	With sorghum leaves
⇓	
Bury jar	Bury up to the neck
⇓	
Mix	Every three days
⇓	
Roll into balls	
⇓	
Sun dry	Three to five days

6.4.2 *Ombolo wa koba*

In Zaire cassava leaves are fermented to produce *ombolo wa koba which* is traditionally eaten with boiled cassava and plantain bananas. Cassava leaves are allowed to wilt and turn black. This takes about three to four days. The cassava leaves are then chopped up and placed in a pot of boiling water for about one hour. During this processing stage, a water soluble extract of ash is produced by placing the ash of burnt dried banana skins and palm tree flowers in a strainer and pouring water through it. The extract is then added to the boiled cassava leaves. The extract is alkaline and neutralises the cyanhydric acid liberated when the leaves are chopped up (Jones *et al,* 1996)[127]. Salt and dried fish or meat is also added. After allowing the cassava leaf mixture to cool a little, acid palm oil is then added. This reacts with the excess alkali and neutralises it. The product is now ready to be eaten (Menea and Bishosha)[128].

CHAPTER 7

PRODUCTS OF MIXED FERMENTATIONS

Most traditional fermented food products are made by a complex interaction of different microorganisms. This chapter deals with the products made when there is not a single dominant set of microorganisms.

7.1 Vinegar

Vinegar is the product of a mixed fermentation of yeast followed by acetic acid bacteria. Vinegar, literally translated as sour wine, is one of the oldest products of fermentation used by man. It is the acetic acid produced by the fermentation of alcohol (ethanol) which gives the characteristic flavour and aroma to vinegar.

It can be made from almost any fermentable carbohydrate source, for example fruits, vegetables, syrups and wine. The basic requirement for vinegar production is a raw material that will undergo an alcoholic fermentation. Apples, pears, grapes, honey, syrups, cereals, hydrolysed starches, beer and wine are all ideal substrates for the production of vinegar. To produce a high quality product it is essential that the raw material is mature, clean and in good condition.

Indigenous vinegars can be made quite simply by the spontaneous fermentation of a fruit or alcohol. All that is necessary is an alcoholic substrate, strains of acetic acid forming bacteria (*acetobacter*) and oxygen to enable the oxidation of alcohol. However, this process is very slow and vinegars produced by this method tend to be of inferior quality. Controlled fermentation conditions produce a more acceptable product. A wide range of raw materials can be made into vinegar.

7.1.1 Coconut water vinegar

Location of production

Throughout Asia particularly the Philippines and Sri Lanka

Product description

A clear liquid with a distinctive acetic acid taste with a hint of a coconut flavour.

Raw material preparation

Coconut water is a waste product, which is produced in appreciable quantities in the Philippines, Sri Lanka Thailand and other countries. Its conversion into vinegar therefore presents an attractive option for decreasing wastage and producing a valuable product.

Processing

Coconut water is a good base for vinegar, but its sugar content is too low (only about 1%). Sugar needs to be added to bring the level of sugar up to 15%. After the addition of sugar, the coconut juice is allowed to ferment for about seven days, during which time the sugar is converted to alcohol. An alternative method is to pasteurise the coconut water and sugar mixture and add yeast.

After this initial fermentation, strong vinegar (10% v/v) is added to stimulate the growth of acetic acid bacteria and discourage further yeast fermentation. The acetic acid fermentation takes approximately one month, yielding vinegar with approximately 6% acetic acid. The fermentation will take less time than this if a generator is used.

After fermentation, the vinegar must be stored in anaerobic conditions to prevent spoilage by the oxidation of acetic acid. (Steinkraus, 1996)[129]

Clarification can be achieved by stirring with a well-beaten egg white, heating until the egg white coagulates and filtering (Anon)[130].

7.1.2 Pineapple peel vinegar

Location of production

Latin America and Asia

Product description

This product enables the utilisation of pineapple peels; which are usually discarded during the processing or consumption of the fruit. The product has a distinct, very light pineapple flavour and has the same uses as any commercial vinegar.

Raw material preparation

The peels should be from very well washed ripe pineapples (damaged, rotten or infected fruits should not be used as a source of peels). Use only the peels, not the leaves or stems. The water used should be potable water, boiled if necessary. All the equipment should be well cleaned, as well as the bottles, which should also be steam-sterilised before use.

Processing

The peels should be cut into thin strips and put into clay or pewter pots. Aluminium or iron pots should not be used.

Sugar and clean water are added. Each pot is then inoculated and covered with a clean cotton cloth, held around the pot with an adhesive tape, to prevent contamination by insects or dust. The inoculated pineapple is fermented at room temperature (about 20-22^{0}C) for about eight days. The acidity should be checked daily. The water level should be maintained during this period. The product should be increasingly acid and by the eighth day it should have the required concentration of 4 per cent acetic acid in vinegar. If higher acidity is desired the product is left to ferment for another one or two days.

The development of acidity should be checked by tasting the product during fermentation.

The residual bacteria removed may be reused as a residue inoculum two or three times more.

The traditional process may be improved by a two-stage fermentation in which alcohol is first formed by yeast (*Saccharomyces cerevisiae*) and the 'must' is then inoculated with acetic acid bacteria (*Acetobacter pasteurianus*). In outline, the process involves liquidising the peels and diluting with water (water: pulp is 4:1), adjusting the pH to 4.0 using sodium bicarbonate and adding yeast nutrient (ammonium phosphate) at 0.14g per litre. A starter culture is added at 2.7g per litre and the fermentation allowed to take place at 25^{0}C for two days. The 'must' is then filtered and inoculated with acetic acid bacteria and allowed to ferment for eleven days with aeration of the 'must'. Other parts of the process are similar. Additional equipment includes a pH meter, refractometer, liquidiser, fermentation locks and equipment for preparing the starter cultures (Fellows, 1997)[131].

Flow Diagram

Pineapple	Well washed in clean water.
⇓	
Peel the fruit	Take care not to damage hands
⇓	
Cut the peel	Cut into thin strips and put into clay or pewter pots.
⇓	
Mix with sugar	Sugar is dissolved in clean water
⇓	
Ferment	Each pot is then inoculated and covered with a clean cotton cloth.
⇓	
Filter	Strain through a cheesecloth
⇓	
Package	

Packaging and storage

The vinegar is bottled in clean glass bottles and stored in a cool dark place.

7.1.3 Palm wine vinegar

Palm wine vinegar is a produced across West Africa. It is a vinegar containing about 4% acetic acid, produced from the oxidation of palm wine. It is mainly consumed by people in urban areas as a salad dressing and meat tenderiser, although it also has medicinal uses and is valued in certain rituals. Palm wine is fermented using the same process as for grape wine vinegar – the oxidation of alcohol to acetic acid. The spontaneous process takes about four days. The optimum fermentation temperature is 30° C.

7.1.4 Coconut toddy vinegar

Coconut toddy vinegar is produced throughout South Asia particularly Sri Lanka. It is a clear liquid with a strong acetic acid flavour and a hint of coconut flavour. The fresh toddy is strained, prior to allowing yeast fermentation to occur naturally for 48 to 72 hours. The yeast cells and debris are then removed by progressive sedimentation. After two to four weeks of settling the fermented toddy is placed in barrels. The alcohol is then converted into acetic acid by acetic acid bacteria, which are naturally present. The process can be hastened by adding vinegar as a starter. The fermented toddy is converted into vinegar in about three months. Ageing for six months, results in a pleasantly flavoured final product (Jayawardena, 1977)[132].

7.1.5 *Nipa* Palm Vinegar

In East Asia particularly Papua New Guinea a vinegar is made from the sap of the *Nipa* palm (*Nypa fruticans*) (Paivoke, Adams and Twiddy, 1984)[133].

7.1.6 Quick process pickles

Quick process pickles are easy to make but do not really constitute a fermented food product. For this technique, vegetables are soaked in a low salt solution for a few hours. They are then drained and placed in a container. The container is filled with a hot vinegar and spice mixture or a hot oil and spice mixture (Kordylas, 1990)[134]. There are hundreds of different recipes utilising locally available fruit and vegetables. For instance the book "Pickles of Bangladesh" has recipes for mango sour pickle, sliced mango pickle, sweet olive pickle, hot olive pickle, sweet tamarind pickle, chalta pickle and green chilli pickle (Azami, 1994)[135].

7.2 <u>Cocoa products</u>

7.2.1 Cocoa powder

Location of production

Africa, Asia and Latin America particularly Cote d'Ivoire, Ghana, Indonesia and Brazil.

Product description

A fine brown powder with the characteristic taste of cocoa. It is a major ingredient in the confectionery and bakery industries. The product has a short shelf life. "Drinking chocolate" is a mixture of cocoa powder and sugar.

Raw material preparation

Cocoa beans are the seeds of the cocoa plant (*Theobroma cacao*). Cocoa pods are cut from the cocoa tree. The pods are cut and the beans removed. Only fully ripe and undamaged beans should be selected. It is important that the beans are processed quickly.

Processing

It was formerly believed that cocoa beans were fermented to remove the adhering pulp (Wood, 1990)[136]. However a good flavour in the final cocoa or chocolate is dependent on good fermentation. Fermentation is carried out in a variety of ways but all depend on heaping a quantity of fresh beans with their pulp and allowing micro-organisms to produce heat (Beckett, 1988)[137]. The majority of beans are fermented in heaps although better results are obtained using boxes, which result in a more even fermentation.

Fermentation lasts from five to six days. During the first day the adhering pulp is liquefied and drains away with the temperature rising steadily. The initial alcoholic fermentation gives way to acetification. This and other chemical changes cause the temperature to rise in excess of 50°C. The beans die. It was thought in the past that death was mainly due to increasing temperature. It is now known that acetic acid at a concentration of 1 percent in the bean is the cause of death and that it is only enhanced by heat, lactic acid and ethanol. The pH value of the cotyledon drops from 6.45 to 4.5 over 120 hours and that during the same period the acetic acid content increased from 0 to 1.36 percent, while the lactic acid content increased from 0.005 to 0.12 percent. When the bean dies maceration of the tissue takes place, allowing enzymes and substrate to mix freely. The possible substrates for enzymes are carbohydrates, lipids, phenolics and amino acids. In addition it is known that the bacteria can metabolise alcohols and organic acids of various kinds. The changing chemical picture is complex. Possible major substrates for microorganisms are carbohydrates, lipids, phenolics and amino acids. Unlike some flavours and aromas, that of chocolate is not attributable to a single compound (Carr, 1985)[138] (Minifie, 1980)[139].

During fermentation the external appearance of the beans changes. At first they are pinkish with a covering of white mucilage. Gradually the colour darkens and the mucilage disappears. The beans on the surface are always darker than those deeper in the heap or box, indicating that the colour change is oxidative. As the beans are mixed, their colour becomes a more uniform orange-brown and they are only slightly sticky. At this stage they are ready for drying.

The beans need to be dried to a moisture content of less than 7.5%. The beans are dried by either being spread out in the sun in layers a few centimetres thick or in artificial dryers. There are numerous types of dryers but it is important that any smoky products of combustion do not come in contact with the beans otherwise taints will appear in the final product.

The beans are cleaned to remove the extraneous matter.

Cocoa beans consist of an outer skin that needs to be removed and inner "nib". The shell is sometimes removed before roasting and sometimes after roasting.

For cocoa powder roasting temperatures of 120 to 150°C are used. There are many designs of roasters: both batch and continuous systems. The operation is controlled so that the cocoa is heated to the required temperature without burning the shell or the cotyledon. The heat is applied evenly over a long period of up to 90 minutes to produce even roasting. The bean must not be contaminated with any combustion products from the fuel used and provision must be made for the escape of any volatile acids, water vapour and decomposition products of the bean (Wood, 1980)[140] (Cakebread, 1975)[141]. After roasting the beans are cooled quickly to prevent scorching.

The roasted nibs are ground into a powder in a plate mill. The resulting powder is sieved through fine silk, nylon or wire mesh.

To produce cocoa powder, some of the cocoa butter needs to be removed. With low fat cocoa powder, more than 90% of the cocoa butter is removed. With medium fat cocoa powder, more than 78% of the cocoa butter has been removed. Finally high fat cocoa powder has less than 78% of the cocoa butter removed. Extrusion, expeller, or screw presses are used in the cocoa industry to remove the cocoa

The cake from the mill is ground in a hammer mill to produce the cocoa powder.

Flow diagram

Cocoa beans	
⇓	
Sort	Select only mature beans
⇓	
Ferment	In heaps or boxes
⇓	
Clean	Remove extraneous material
⇓	
Dry	In the sun or in artificial dryers
⇓	
Roast	120 to 150°C
⇓	
Grind	
⇓	
Remove cocoa butter	
⇓	
Grind cake	
⇓	
Pack	

Packaging and storage

Cocoa powder is hygroscopic (picks up moisture from the air) and should be protected, especially in humid climates. Lidded tins or sealed polythene bags should be used.

7.2.1 Chocolate

Location of production

Throughout Africa, Asia and Latin America.

Product description

A brown solid oily product with the characteristic taste of chocolate. It is a major ingredient in the confectionery and bakery industries.

Preparation of raw materials

Cocoa beans are the seeds of the cocoa plant (*Theobroma cacao*). Cocoa pods are cut from the trees, and the beans are removed from the pods. Only fully ripe and undamaged beans should be selected. It is important that the beans are processed quickly.

Processing

Fermentation, drying and cleaning of the beans have been described in Section 7.2.1. For cocoa butter production the roasting temperatures are 100°C to 104°C There are many designs of roasters: both batch and continuous systems. The operation is controlled so that the cocoa is heated to the required temperature without burning the shell or the cotyledon. The heat is applied evenly over a long period of up to 90 minutes to produce even roasting; the nib must not be contaminated with any combustion products from the fuel used and provision must be made for the escape of any volatile acids, water vapour and decomposition products of the nib (Wood, 1980)[142]. After roasting the beans are cooled quickly to prevent scorching

Roasting will have already loosened the shell. The beans are then lightly crushed with the object of preserving large pieces of shell and nib and avoiding the creation of small particles and dust. The cocoa bean without its shell is known as a "cocoa nib". The valuable part of the cocoa bean is the nib, the outer shell being a waste material of little value.

Alkalisation is a treatment that is sometimes used before and sometimes after grinding to modify the colour and flavour of the product. This was developed in the Netherlands in the last century and is sometimes known as "Dutching". This involves soaking the nib or the cocoa mass in potassium or sodium carbonate. By varying the ratio of alkali to nib, a wide range of colours of cocoa powder can be produced (Glossop, 1993)[143]. Complete nib penetration may take an hour. After alkalisation the cocoa needs to be dried slowly.

The cocoa nib is ground into "cocoa liquor" (also known as "unsweetened chocolate" or "cocoa mass"). The grinding process generates heat and the dry granular consistency of the nib is turned into a liquid as the high amount of fat contained in the nib melts (Gates, 1990)[144].

There are various pre-treatments to develop the flavour of the cocoa mass with and without reaction solutions. These include the "Luwa thin-layer evaporator", "Petzomat thin-layer process", "Cocovap process", "Lehman KFA process" and "Carle-Montanari process" (Beckett, 1988) [145].

Extrusion, expeller, or screw presses are used in the cocoa industry for the production of cocoa butter from whole beans, and mixtures of fine nib dusts, small nibs, and immature beans. Research in the Kerala Agricultural University has led to develop a suitable pressure device capable of separating cocoa butter from ground cocoa mass ideally suitable for small scale manufacturers (Ganeshan, 1990)[146]. In Peru a simple screw press is used to extract cocoa butter from beans. The crude cocoa butter is filtered through cloth and allowed to solidify.

To produce plain chocolate, cocoa mass is mixed with sugar and sufficient cocoa butter to enable the chocolate to be moulded. The ratio of mass to sugar varies according to the national taste. The mixture is ground to such a degree that the chocolate is smooth to the palate. At one time this was done by a lengthy process in melengeurs - heavy granite rollers in a revolving granite bed - but nowadays grinding is done in a series of rolls.

The chocolate is then "Conched". This may last for several hours. The chocolate is heated, this helps to drive off volatile acids, thereby reducing acidity when present in the raw bean, and the process finishes the development of flavour and makes the chocolate homogeneous (Wood, 1980) [147]. Similar processes are involved in the manufacture of milk chocolate. The milk is added in various ways either in powder form to the mixture of mass, sugar and cocoa butter, or by condensing first with sugar, adding the mass and drying this mixture under vacuum. The product is called 'crumb' and this is ground and conched in a similar manner to plain chocolate. After conching the chocolate has to be tempered before it is used for moulding or for enrobing confectionery centres. Tempering involves cooling and reaching the right physical state for rapid setting after moulding or enrobing.

7.3 Coffee

Location of production

Throughout Africa, Asia and Latin America particularly Brazil, Colombia, Indonesia, Mexico and Cote d'Ivoire.

Product description

A fine dark brown powder made from roasted coffee beans. Brewed with boiling water and consumed as a drink.

Raw material preparation

Coffee beans are harvested from two plants *Coffea arabica and Coffea canephora* variety *robusta*. Only ripe berries should be used in coffee production. Berries can be placed in water so those immature berries which float can be identified and discarded.

Processing

Dry processing is the simpler of the two processing methods and is popular in Brazil for the processing of *Robusta* coffee and in Sri Lanka for processing *Arabica* coffee. The coffee cherries are dried immediately after harvest by sun drying on a clean dry floor or on mats. The dried berry is then hulled to remove the pericarp. This can be done by hand using a pestle and mortar or in a mechanical huller. The mechanical hullers usually consist of a steel screw, the pitch of which increases as it approaches the outlet so removing the pericarp. The hulled coffee is cleaned by winnowing.

Wet processing involves squeezing the berry in a pulping machine or pounding in a pestle and mortar to remove the outer fleshy material (mesocarp and exocarp) and leave the bean covered in mucilage. This mucilage is removed by fermentation. Fermentation involves placing the beans in plastic buckets or tanks and allowing them to sit, until the mucilage is broken down. Natural enzymes in the mucilage and yeasts and bacteria in the environment, work together to break down the mucilage. The coffee should be stirred occasionally and every so often a handful of beans should be tested by washing in water. If the mucilage can be washed off and the beans feel gritty rather than slippery, the beans are ready.

There is much debate about the fermentation of coffee beans. Some researchers feel that the mucilage breakdown is caused by enzymatic breakdown. For instance Wellman has stated that enymatic fermentation starts immediately the beans have been squeezed from the fresh berries. If these 'pulped' beans are piled up or put in a container and protected from any bacterial or other contamination the fermentation will progress. After a number of hours the enzymes of the pulp will have acted on the torn tissues, gorged with starches, sugars and pectins, in such a manner that, without any microbial intervention, the remaining pulp will be easily detached from the beans and washed off in water (Wellman, 1961)[148]. However most investigators acknowledge the necessity for the presence of microorganisms for the depectinisation of the beans.

The following micro-organisms have been isolated: *Leuconostoc mesenteroides; Lactobacillus plantarum; Lactobacillus brevis, Streptococcus faecalis, Aerobacter (Enterobacter)* and *Escherichia,* pectinolytic species of *Bacillus, Saccharomyces marscianus, S. bayanus and* a *Flavobacterium* sp., *Erwinia dissolvens, Fusarium spp, Aspergillus spp* and *Penicillium. (Pedersen and Breed)*[149] (Vaughn *et al., 1*958)[150] (Hilmer *et al.,* 1965)[151], (Agate and Bhat, 1966)[152]

The beans should then be washed immediately as 'off' flavours develop quickly. To prevent cracking the coffee beans should be dried slowly to 10% moisture content (wet basis). Drying should take place immediately after to prevent 'off' flavours developing. The same drying methods can be used for this as for the dry processed coffee. After drying the coffee should be rested for 8 hours in a well-ventilated place. The thin parchment around the coffee is removed either by hand, in a pestle and mortar or in a small huller. The hulled coffee is cleaned by winnowing.

The final flavour of the coffee is heavily dependent on how the beans are roasted. Roasting is a time temperature dependent process. The roasting temperature needs to be about 200°C. The degree of roast is usually assessed visually. One method is to watch the thin white line between the two sides of the bean, when this starts to go brown the coffee is ready. As preferences vary considerably from region to region, a lot of research will need to be done to find the locally acceptable degree of roast. Coffee beans can be roasted in a saucepan as long as they are continually stirred. Roasting the coffee in sand makes a small improvement, as this provides a more even heat. A roaster will produce a higher quality product.

Grinding is a means of adding value to a product. However, it is fraught with difficulties. It is easy to make an assessment of an intact bean, while a ground product presents some difficulty. The fear of adulteration and the use of low quality produce are justified. Because of this there is a great deal of market resistance to ground coffee. This market resistance can only be overcome by consistently producing a good product. There are basically two types of grinders - manual grinders and motorised grinders (anon, 1995)[153].

Flow diagram

Coffee cherry	
⇓	
Sort the cherries	Select only mature undamaged beans
⇓	
Pulp	Skin and pulp removed
⇓	
Ferment	To remove mucilage
⇓	
Dry the beans	Sun drying or artificial drying
⇓	
Roast	
⇓	
Grind	
⇓	
Package	

Packaging and storage

Roasted beans can be stored in sacks. Milled beans need to be packaged quickly to prevent the loss of volatile flavour components. The packaging material should be airtight. Polythene is not suitable as it is a low barrier to loss of aroma (Fellows, 1997)[154].

7.4 Other mixed fermentation products

7.4.1 Vanilla

Vanilla is produced in Madagascar, Indonesia and various South Pacific islands. It is a dark brown pod about 20 cm in length. Vanilla is produced by fermenting the pods of the orchids of the genus *vanilla*. The pods are first sun dried for 24 to 36 hours and then blanched in hot water (65°C) for two to three minutes. The pods are then fermented in boxes and dried again.

7.4.2 Tabasco

Tabasco sauce is made in Mexico and Guatemala. The chilli pods are harvested, ground into a paste and placed in a container with salt. The hot and fiery sauce develops.

7.4.3 Tea

In the production of tea, there is a process referred to as fermentation. However microbial activity is not involved in the so-called 'fermentation' of tea. The chemical changes are effected by enzymes alone. Fermentation rooms are used where moisture and temperature can be controlled. During fermentation even further darkening of the leaf occurs and the typical aroma develops. By subjective judgement of the aroma's intensity the period necessary for completion is gauged (Carr, 1985)[155].

CHAPTER 8

THE WAY AHEAD

Fermentation is one of the oldest forms of food preservation technology in the world. Indigenous fermented foods such as bread, cheese and wine, have been prepared and consumed for thousands of years and are strongly linked to culture and tradition, especially in rural households and village communities. Fermented foods are popular throughout the world and in some regions make a significant contribution to the diet of millions of individuals. For instance Soy sauce is consumed throughout the world and is a fundamental ingredient in diets from Indonesia to Japan. Over one billion litres of soy sauce are produced each year in Japan alone. In Africa fermented cassava products (like *gari* and *fufu*) are a major component of the diet of more than 800 million people and in some areas these products constitute over 50% of the diet. Fermentation is a relatively efficient, low energy preservation process, which increases the shelf life and decreases the need for refrigeration or other forms of food preservation technology. It is therefore a highly appropriate technique for use in developing countries and remote areas where access to sophisticated equipment is limited. There is tremendous scope and potential for the use of micro-organisms towards meeting the growing world demand for food, through efficient utilisation of available natural food and feed stocks and the transformation of waste materials.

There is a danger that the introduction of 'western foods' with their glamorous image will displace these traditional fermented foods. Although fermentation of foods has been in use for thousands of years for the preservation and improvement of a range of foods, the microbial and enzymatic processes responsible for the transformations were, and still are, largely unknown. Because of the tremendously important role indigenous fermented fruits and vegetables play in food preservation and their potential to contribute to the growing food needs of the world, it is essential that the knowledge of their production is not lost. Moreover, it is essential to increase the knowledge and understanding of the methods of preparation, in order to improve the efficiency of fermentation, especially the traditional processes as the yields of traditional fermentation processes are often low and sometimes the products are unsafe. This chapter discusses potential areas for improvement of indigenous fermented fruit and vegetable products. It can be divided into four main areas:

- Improve the understanding of the fermented products
- Refine the processes
- Disseminate the improvements
- Create a supportive policy environment.

8.1 Improving the understanding of fermented products

For fermented products such as cheese, bread, beer and wine, which are produced on a commercial scale, a good understanding of the microbial processes has been developed. However with many of the fermented products in Africa, Asia and Latin America, knowledge of the processes involved is poor. It is likely that the basic principles apply across the board, but production conditions vary enormously from region to region, giving rise to numerous variations of the basic fermented product. It is not the intention or the desire to standardise the process and thereby lose this huge diversity, rather it is to harness the tremendous potential these methods have to contribute to increasing not only the quantity, but quality of food available to the worlds population.

There are two main reasons for gaining a better understanding of indigenous fermented products:

- Documenting the traditional knowledge to ensure that the huge diversity is not lost
- Developing a scientific understanding of the microbial processes, with a view to improving the efficiency of the process

8.1.1 Documenting the traditional knowledge

Fermentation is one of the oldest food processing technologies in the world. The knowledge of how to make these products has often been passed down from parent to child (usually mother to daughter) and belongs to that undervalued body of "indigenous knowledge". Most of this knowledge has not been documented and is in danger of being lost as technologies evolve and families move away from traditional food preservation practices.

The collection and preservation of indigenous knowledge is of interest to governments, historians, anthropologists and scientists, to name but a few. Several individuals and organisations are actively involved in research in this area:

Several research institutes and scientists in Africa, Asia and Latin America are recording information on traditional fermented foods. Dr Karki in Nepal and Hamid Dirar have been extensively quoted in this book and their work on fermented products in Nepal and Sudan is of great importance.

Intermediate Technology has been collecting information about traditional food products from Africa, Asia and Latin America. The first volume of products was published in 1997 (Fellows, 1997)[156]. This included the following fermented food products: *kenkey*, *ogi*, *injera*, fermented sweet bread, sorghum beer, palm wine, banana beer, pineapple peel vinegar, lime pickle, tamarind pickle, mango pickle, *gari*, vegetable pickle, coconut toddy, yoghurt, *ayib* and *dawdawa*. Volume 2 will be published in 1999 and regional publications on the traditional food products of Bangladesh, Southern Africa and the Andes are planned.

The European Union has just completed an audit of the traditional food products of Europe. The results have been published in a series of publications (Cowan, 1998)[157].

The Special Programme on Biotechnology and Development Cooperation for the Netherlands Government was established in 1992 to improve the access of developing countries to biotechnological expertise and innovation with a focus on using biotechnology for the benefit of small-scale farmers and producers. Part of the programme involved a competition to identify farmers' existing biotechnology practises. This programme resulted in the collection of valuable information about traditional fermented food products (Bunders *et al*, 1996)[158].

Finally, the Food and Agriculture Organisation of the United Nations sees the value in collecting and preserving this source of knowledge. This book is an example of their work.

8.1.2 Developing a scientific understanding of the microbial processes

Most traditional fermented products are made by natural fermentations carried out in a non-sterile environment. The specific environmental conditions cause a gradual selection of microorganisms responsible for the desired final product. This is appropriate for small-scale production for home consumption. However the method is difficult to control and there are risks of accompanying micro-flora causing spoilage and unsafe products.

If the processes are to be refined, with a view to production on a larger scale, it is essential to have a scientific understanding of the fermentation processes. This can be developed by:

- The isolation and characterisation of the essential micro-organisms involved;
- The determination of the role of external factors in fermentations and the effects of these on the metabolism of micro-organisms
- The investigation of the effects of pre-treatments of raw materials on the fermentation process
- The identification of the options for further processing and how these affect the taste and texture of the product.

This research is capital intensive and usually requires scarce foreign exchange. It requires the use of sophisticated equipment and reagents backed with a consistent energy and water supply, which are not always available in developing countries. To meet the current and future challenges in developing countries, it is important that these countries develop the capabilities to benefit from improvements in fermentation methodologies.

Biotechnologies need to be developed which are affordable by the poor, since it is they who are likely to benefit most by improvements to the traditional processes.

8.2 Refining the process

The art of traditional processes needs to be refined to incorporate objective methods of process control and to standardise quality of the final product without losing their desirable attributes such as improved keeping quality, taste and nutritional qualities.

Science based fermentation research has often focused mainly on improving the metabolic properties of the micro-organisms used as a starter culture, using the techniques of selection, mutation and genetic modification. To achieve this a full understanding of the fermentation process is needed. The properties of the starter culture must be known and culture conditions must be controllable. This means that science based fermentation research currently has little to offer traditional food processing (Broerse and Visser, 1996)[159].

8.2.1 Process control

Once the details of the fermentation process and the microbes involved are known and understood, it is possible to begin to refine and improve the process. The commercially produced products, such as bread, wine, soy sauce and pickles are all examples of processes, which have been studied, and optimised. Recently traditional fermented products in Africa have been industrialised. In Nigeria, Dadwa (a fermented legume product) is now made by Cadburys and in Zimbabwe, traditional fermented milk is made industrially and sold as "Lacto" (Okafor, 1992)[160].

The areas where the efficiency and yield of food fermentation processes can be increased are:

- The selection or development of more productive microbial strains
- The control and manipulation of culture conditions
- The improvement of product purification and concentration.

Techniques to control the fermentation processes could include the development of pure starter cultures. Developing these by laboratory selection or genetic engineering is not viable. A more feasible approach would be to exploit the ecological principle of inoculum enrichment by natural selection. Another approach to stabilise fermentation under nonsterile conditions is the use of multi-strain dehydrated starters which can be stored at ambient temperature. These are already used for the manufacture of tempeh (Nout, 1992)[161].

8.2.2 Quality control

The aim of quality control is to ensure that every batch of food produced has a satisfactory and uniform quality. This does not necessarily mean that it is the highest quality possible but that it reaches the standard the customers are willing to pay for (Fellows, 1996)[162].

Inadequate quality control can have an adverse effect on local demand for the product. This is particularly a problem for small-scale traditional production. In modern industrial applications, the fermentation equipment and processes are controlled using expensive technology, resulting in a consistent product of a known quality. Traditional practises take place in a less predictable environment. This can result in mistakes including sour beer and mouldy pickles.

It is often felt that traditional products made at the small scale are unhygienic and unsafe. This is sometimes true. However the case is often overstated. Many fermented foods are inherently safe due to low moisture contents or high acidity. Lime pickle from India and Gundruk from Nepal are good examples. Several of the steps in traditional processing are designed to reduce contamination. These include boiling, adding salt and sun drying.

Quality control procedures are essential for the production of safe products and contribute to the success of small food processing businesses. Appropriate quality control procedures need to be developed and implemented. These procedures need to be developed with the processors who must understand and apply them. The quality of food is highly subjective. What is acceptable to one customer is not acceptable to another. It is important to carry out participatory research to identify ways to improve the quality control procedures for fermented food products.

The sort of areas that should be investigated include:

- Selecting good quality raw materials
- Processing under correct conditions
- Ensuring high standards of personal hygiene by the food processors
- Ensuring the processing area is sufficiently clean
- Using correct packaging

8.3 Disseminating improvements

Documentation of the traditional methods of food fermentation and research to identify improved methods of production are meaningless if the results are not disseminated to those who are likely to put them into practice.

There is a danger of mystifying the fermentation process by enrobing it in scientific theory. What was once a simple process carried out by any family member, in the confines of the household, using locally available equipment and materials, could become a process to be feared. It is important to be realistic and to ensure that the improvements recommended are ones, which can easily be put into practice. The aim is not to deter the production of fermented foods at the small scale, but to encourage their production and consumption on a larger scale.

Fermented foods often have a stigma attached to them – they are considered as poor mans food. As soon as a family can afford to buy processed foods, they move away from carrying out home fermentation. This is a pity, because as we have seen earlier, fermented food products have many nutritional advantages, which surpass western-style fast foods and processed foods. Where cultural values attached to the fermented food are strong, it is unlikely that there will be an image problem. For example, kimchi is considered part of the national heritage in Korea. It is a vital ingredient of all meals and as such is a highly valued food. This is reflected in the amount of research carried out on the product and the detailed understanding of the process, which already exists.

It is not difficult to gain access to village people, both to collect the traditional information and to disseminate improved practices. Numerous organisations are involved in field projects. They can be used to organise training sessions and group meetings for the dissemination of new methods, for example, for the use of pure starter cultures.

One of the problems likely to be encountered is gaining access to starter cultures and other improved methods. Agricultural extension services should take a responsibility for the promotion and the supply of starter cultures at a price, which is affordable.

8.4 Creating a supportive environment for production of fermented food products

Developing countries need to build their resources of trained, knowledgeable individuals, who are able to apply the basic microbiological principles to the production of fermented foods. Extra support should be made available to train professionals in this discipline.

Fermented foods should be recognised as part of each countries heritage and culture and efforts made to preserve the methods of production. A recognised body (government or non-government) should take the responsibility for the collection of details and the promotion of fermented food products. Consumers need to be made aware of the numerous benefits of fermented foods and their prejudices against fermented foods; especially those traditionally produced at the home scale, dispelled.

Many traditional fermented foods are produced from minor or wild fruits and vegetables, many of which are being lost through deforestation and loss of biodiversity.

References

[1] Stanton, R.W., (1985), *Food Fermentation in the Tropics*, in "Microbiology of Fermented Foods", edited by Wood, B.J.B., Elsevier Applied Science Publishers, UK.

[2] Dirar, H., (1993), *The Indigenous Fermented Foods of the Sudan: A Study in African Food and Nutrition*, CAB International, UK

[3] Pedersen, C.S., (1979), *Microbiology of Food Fermentations*, AVI Publishers, USA

[4] Sugihara, T.F., (1985), *Microbiology of Breadmaking*, in "."Microbiology of Fermented Foods", edited by Wood, B.J.B., Elsevier Applied Science Publishers, UK

[5] Yokotsuka, T., (1985), *Fermented Protein Foods in the Orient, with Emphasis on Shoyu and Miso in Japan*, in "Microbiology of Fermented Foods", edited by Wood, B.J.B., Elsevier Applied Science Publishers, UK

[6] Walker, P.M.B. (1988), *Chambers Science and Technology Dictionary*, Chambers, Cambridge University Press, UK.

[7] Little, W., Fowler, H.W. and Coulson J., (1973), *The Shorter Oxford English Dictionary*, Oxford University Press, UK.

[8] Walker, P.M.B. (1988), *Chambers Science and Technology Dictionary*, Chambers, Cambridge University Press, UK.

[9] Anon, (1993). *Fruit and Vegetable Processing*. Food Cycle Technology Source Books, Intermediate Technology Publications Ltd, UK.

[10] Little, W., Fowler, H.W. and Coulson, J., (1973), *The Shorter Oxford English Dictionary*, Oxford University Press, UK.

[11] Anon, (1993). *Fruit and Vegetable Processing*. Food Cycle Technology Source Books, Intermediate Technology Publications Ltd, UK.

[12] Anon, (1992) *Applications of Biotechnology to Traditional Fermented Foods*, report of an Ad Hoc Panel of the Board on Science and Technology for International Development, (1992), National Academy Press, Washington D.C., USA.

[13] Anon, (1996), *The Budapest Declaration*, World Congress of Food Science and Technology, Budapest, Hungary

[14] Anon, (1995), Development Education Exchange Paper, FAO, Rome

[15] Anon, (1996), *Food Processing for Food Security and Income Generation*, Intermediate Technology, UK

[16] Dirar, H., (1992), *Sudan's Fermented Food Heritage*, in "Applications of Biotechnology to Traditional Fermented Foods", National Academy Press, USA

[17] Arthur, R.A.J., (1986), *Tribal Recipe may help to Feed the World*, London Press Service 060416, UK

[18] Karki,T., (1986), *Some Nepalese Fermented Foods and Beverages*, in "Traditional Foods: Some Products and Technologies", Central Food Technological Research Institute, India

[19] Dirar, H., (1993), *The Indigenous Fermented Foods of the Sudan: A study in African Food and Nutrition*, CAB International, UK

[20] Steinkraus K H, (1996). *Handbook of Indigenous Fermented Foods*. Marcel Decker Inc, New York.

[21] Anon, (1995), *Food for Consumers*, Food and Agriculture Organisation, Rome, Italy

[22] Anon, (1989), *Sub Saharan Africa*, World Bank, Washington

[23] Conroy, C., Gordon, A. and Marter, A., (1995), *Development and Dissemination of Agro Processing Technologies*, Natural Resources Institute, UK

[24] Oyewole, O.B. (1992), *Cassava Processing in Africa*, , in "Applications of Biotechnology to Traditional Fermented Foods", Report of an Ad Hoc Panel of the Board on Science and Technology for International Development, National Academy Press, Washington D.C., USA

[25] Anon, (1982), *Production Yearbook*, FAO, Italy

[26] Steinkraus, K.H. (1992), *Lactic Acid Fermentations*, in "Applications of Biotechnology to Traditional Fermented Foods", report of an Ad Hoc Panel of the Board on Science and Technology for International Development, National Academy Press, Washington D.C., USA

[27] Kovac, B, (1997), *The Use of the Mould Rhizopus oligosporus in Food Production*, Food Technology and Biology

[28] Parades-Lopez, O., (1992), *Nutrition and Safety Considerations in Applications of Biotechnology to Traditional Fermented Foods*, report of an Ad Hoc Panel of the Board on Science and Technology for International Development, National Academy Press, Washington D.C., USA

[29] Dirar, H., (1992), *Sudan's Fermented Food Heritage*, in "Applications of Biotechnology to Traditional Fermented Foods", National Academy Press, USA.

[30] Hammes, W.P. and Tichaczek, P.S., (1994), *The Potential of Lactic-acid Bacteria for the Production of Safe and Wholesome Food,* Zeitschrift fur Lebenmitteltechnol, Germany

[31] Wood, B.J.B. and Hodge, M.M., (1985), *Yeast Lactic Acid Bacteria Interaction*, in "Microbiology of Fermented Foods", Edited by Wood, B.J.B., Elsevier Applied Science Publishers, UK

[32] Matsusaki, H., Sonomoto, K. and Ishizaki, A., (1997*), Bacteriocins, Growth Inhibitory Substances of Lactic Acid Bacteria,* Seibutsu Kogaku Kaishu Journal of the Soceity for Fermentation and Bioengieering, Japan

[33] Adams, M.R. and Nicolaides, L.,(1997), *Review of the Sensitivity of Different Foodborne Pathogens to Fermentation*, Food Control, UK

[34] Gorama, H. and Bullerman, L.B., (1995*), Anitmycotic and Antiaflatoxigenic Effect of Lactic Acid Bacteria* - a Review, Journal of Food Protection.

[35] Nout, M.J.R., (1995), *Fungal Interactions in Food Fermentations*, Canadian Journal of Botany, Canada.

[36] Ottogalli, G. and Galli, A., (1997), *Fermented Foods in the Past and in the Future*, Annali di Microbiologia ed Enzimologia, Italy

[37] Motarjemi, Y., Nout, M.J.R., Adams, M., Bosman, L., Fuchs, R., Harcourt, D., Hastings, J.W., Von Holy, J.W., Holzapfel, A., Kouthon, G., Lee, C.H. and Liebenerg, (1996*), Food Fermentation: a Safety and Nutrtional Assessment*, Bulletin of the World Health Organisation, Switzerland.

[38] Frohlich, R.H., Kunze, M. and Kiefer, I., (1997), Cancer Preventive Impact of Naturally Occurring, Non-nutritive Constituents in Food, Acta Medica Austriaca, Austria

[39] Watson, F.E., Ngesa, A., Onyango, J., Alnwick, D. and Tomkins, A.M., (1996), Fermentation - A traditional Anti-Diarrheal Practice Lost, International Journal of Food Sciences and Nutrition.

[40] Svanberg, B., (1992), *Fermentation of Cereals: Traditional Household Technology with Nutritional Benefits for Young Children*, IDRC Currents 2, Canada

[41] Battcock, M.J., (1992), *Street Foods and Development in Bangladesh*, Food Laboratory News, Volume 8:2, Germany

[42] Cambell Platt, G., (1987). *Fermented Foods of the World – A Dictionary And Guide.* Butterworths, London, UK.

[43] Mountney, G. J. and Gould, W. A. (1988). *Practical Food Microbiology and Technology.* AVI Books, Van Nostrand Reinhold Company, New York, USA.

[44] Mountney, G. J. and Gould, W. A. (1988). *Practical Food Microbiology and Technology.* AVI Books, Van Nostrand Reinhold Company, New York, USA.

[45] Fellows, P. (1988). *Food Processing Technology: Principles and Practice.* VCH Publishers, UK.

[46] Mountney, G. J. and Gould, W. A. (1988). *Practical Food Microbiology and Technology.* AVI Books, Van Nostrand Reinhold Company, New York, USA.

[47]Walker, P. M. B. (1988). *Chambers Science and Technology Dictionary*, Chambers, Cambridge University Press, UK.

[48]Adams, M. R. and Moss, M. O., (1995). *Food Microbiology*. The Royal Society of Chemistry, Cambridge, UK.

[49]Mountney, G. J. and Gould, W. A. (1988). *Practical Food Microbiology and Technology*. AVI Books, Van Nostrand Reinhold Company, New York, USA.

[50]Board, R. G., (1983). *A Modern Introduction to Food Microbiology*. Blackwell Scientific Publications, Oxford, UK.

[51] Fleet, G. H., (1998). *The microbiology of alcoholic beverages*. In: "Microbiology of Fermented Foods", Blackie Academic and Professional, London, UK.

[52]Stanier, R.Y, Doudoroff, M. and Adelberg, E.A., (1972), *General Microbiology*, Macmillan, UK

[53]Adams, M.R., (1985), *Vinegar*, in "Microbiology of Fermented Foods", Ed. Woods, B.J.B., Elsevier, UK

[54]Stanier, R.Y, Doudoroff, M. and Adelberg, E.A., (1972), *General Microbiology*, Macmillan, UK

[55]Stanier, R.Y, Doudoroff, M. And Adelberg, E.A., (1972), *General Micriobiology*, Macmillan, UK

[56] Fleet, G. H., (1998). *The microbiology of alcoholic beverages*. In: "Microbiology of Fermented Foods", Blackie Academic and Professional, London, UK.

[57] Fleet, G. H., (1998). *The microbiology of alcoholic beverages*. In: "Microbiology of Fermented Foods", Blackie Academic and Professional, London, UK. UK

[57] Fleet, G. H., (1998). *The microbiology of alcoholic beverages*. In: "Microbiology of Fermented Foods", Blackie Academic and Professional, London, UK.

[57] Fleet, G. H., (1998). *The microbiology of alcoholic beverages*. In: "Microbiology of Fermented Foods", Blackie Academic and Professional, London, UK.

[57] Fleet, G. H., (1998). *The microbiology of alcoholic beverages*. In: "Microbiology of Fermented Foods", Blackie Academic and Professional, London, UK.

[58] Fleet, G. H., (1998). *The microbiology of alcoholic beverages*. In: "Microbiology of Fermented Foods", Blackie Academic and Professional, London, UK.

[59] Fleet, G. H., (1998). *The microbiology of alcoholic beverages*. In: "Microbiology of Fermented Foods", Blackie Academic and Professional, London, UK.

[60] Ranken, M.D., Kill, R.C. and Baker, C.G.J., (1997), Food Industries Manual, Blackie Academic and Professional, UK

[61] Fleet G H, (1998). *The microbiology of alcoholic beverages*. In: *Microbiology of Fermented Foods*, Blackie Academic and Professional, London, UK.

[62] Rhodes, A. and Fletcher, D., (1966), *Principles of Industrial Microbiology*, Pergamon Press, UK

[63] Colquichagua, D., (1994), *Vino de Fruta*, ITDG, Lima, Peru

[64] Ranken, M.D., Kill, R.C. and Baker, C.G.J., (1997), Food Industries Manual, Blackie Academic and Professional, UK

[65] Ranken, M.D., Kill, R.C. and Baker, C.G.J., (1997), Food Industries Manual, Blackie Academic and Professional, UK

[66] Fleet G H, (1998). *The microbiology of alcoholic beverages*. In: *Microbiology of Fermented Foods*, Blackie Academic and Professional, London, UK.

[67] Fellows, P., (1997), *Traditional Foods*, Intermediate Technology Publications, UK

[68] Davies, G., (1994), Domestic Banana Beer Production in Mpigi District Uganda, ETC paper, Netherlands

[69] Wimalsiri, P., Sinnatamby, A., Samaranayake, S and Samarasinghr, C.R., (1971), *Cashew Apple Wine,* Industry Prospect Report 44, Industrial Development Board, Sri Lanka,

[70] Aidoo, K.E., (1986), *Lesser-known fermented plant foods*, Tropical Science, 1986, 26, 249-258

[71] Rose, A., (1961), *Industrial Microbiology*, Butterworths, UK

[72] Dirar, H., (1993), *The Indigenous Fermented Foods of the Sudan: A study in African Food and Nutrition*, CAB International, UK

[73] Rose, A., (1961), *Industrial Microbiology*, Butterworths, UK

[74] Steinkraus K H, (1996). *Handbook of Indigenous Fermented Foods*. Marcel Decker Inc, New York.

[75] Odunfa, S.A., (1985), *African Fermented Foods*, in "Microbiology of Fermented Foods", Elsevier Applied Science Publishers, USA,

[76] Fellows, P., (1997), *Traditional Foods*, Intermediate Technology Publications, UK

[77] Odunfa, S.A., (1985), *African Fermented Foods*, in "Microbiology of Fermented Foods", Elsevier Applied Science Publishers, UK.

[78] Fellows, P., (1997), *Traditional Foods*, Intermediate Technology Publications, UK

[79] Wood, B.J.B. and Hodge, M.H., (1985), *Yeast-Lactic Acid Bacteria Interactions and their Contribution to Fermented Foodstuffs*, in "Microbiology of Fermented Foods", Edited by Wood, B.J.B., Elsevier Applied Science Publishers.

[80] Steinkraus, K., (1992), *Lactic Acid Fermentations* in "Applications of Biotechnology to Traditional Fermented Foods", National Academy Press, USA

[81] Steinkraus K H, (1996). *Handbook of Indigenous Fermented Foods*. Marcel Decker Inc, New York.

[82] Steinkraus K H, (1996). *Handbook of Indigenous Fermented Foods*. Marcel Decker Inc, New York.

[83]Walker, P.M.B. (1988), *Chambers Science and Technology Dictionary*, Chambers, Cambridge University Press, UK

[84] Axelsson, L., (1998). *Lactic Acid Bacteria: Classification and Physiology*. In: Lactic Acid Bacteria, Microbiology and functional Aspects. Ed. S Salminen and A von Wright, Marcel Decker Inc, New York, USA.

[85] Axelsson, L., (1998). *Lactic Acid Bacteria: Classification and Physiology*. In: Lactic Acid Bacteria, Microbiology and functional Aspects. Ed. S Salminen and A von Wright, Marcel Decker Inc, New York, USA.

[86] Beuchat, L. R., (1995). *Application of biotechnology to fermented foods*. Food Technology, Jan 1995, 97-99.

[87] Steinkraus K H, (1996). *Handbook of Indigenous Fermented Foods*. Marcel Decker, New York.

[88] Lal, G., Siddappa, G.S. and Tandon, G.L. ,(1986), *Preservation of Fruits and Vegetables*, Indian Council of Agricultural Research, India.

[89] Lal, G., Siddappa, G.S. and Tandon, G.L. ,(1986), *Preservation of Fruits and Vegetables*, Indian Council of Agricultural Research, India.

[90] Kordylas, J.M., (1990), *Processing and Preservation of Tropical and Subtropical Foods*, Macmillan, UK

[91] Pederson, C. S., (1979). *Microbiology of Food Fermentations*. AVI Publishers, USA

[92]Vaughn, R.H., (1985), *The Microbiology of Vegetable Fermentation*, in "Microbiology of Fermented Foods", Wood, B.J.B., Elsevier, UK

[93] Steinkraus, K. H., (1996). *Handbook of Indigenous Fermented Foods*. Marcel Decker Inc, New York, USA.

---

[94] Steinkraus, K. H., (1996). *Handbook of Indigenous Fermented Foods*. Marcel Decker Inc, New York, USA.

[95] Lal, G., Siddappa, G.S. and Tandon, G.L. ,(1986), Preservation of Fruits and Vegetables, Indian Council of Agricultural Research, India

[96] Lal, G., Siddappa, G.S. and Tandon, G.L. ,(1986*), Preservation of Fruits and Vegetables,* Indian Council of Agricultural Research, India

[97] Kordylas, J.M., (1990*), Processing and Preservation of Tropical and Subtropical Foods,* Macmillan, UK

[98]Fellows, P., (1997), *Traditional Foods*, Intermediate Technology Publications, UK

[99] Karki,T., (1986), *Some Nepalese Fermented Foods and Beverages*, in "Traditional Foods: Some Products and Technologies", Central Food Technological Research Insitute, India

[100]Vaughn, R.H., (1985), *The Microbiology of Vegetable Fermentation*, in "Microbiology of Fermented Foods", Wood, B.J.B., Elsevier, UK

[101] Pedersen, C.S., (1979), *Microbiology of Food Fermentation*, AVI Publishing Co, USA

[102]Boon-Lung, N., (1986), *Traditional Technologies of Thailand: Traditional Fermented Food Products*, in Traditional Foods: Some Products and Technologies, Central Food Technological Institute, India.

[103] Steinkraus K H, (1996). *Handbook of Indigenous Fermented Foods*. Marcel Decker Inc, New York.

[104] Pedersen, C.S., (1979), *Microbiology of Food Fermentation*, AVI Publishing Limited, USA

[105] Lal, G., Siddappa, G.S. and Tandon, G.L. ,(1986), Preservation of Fruits and Vegetables, Indian Council of Agricultural Research, India

[106]Redelinghuys, H. and Van der Riet, W., (1978) , *Green Mango Aachar*, National Food Research Institute, Council for Scientific and Industrial Research, Pretoria, South Africa.

[107]Redelinghuys, H. and Van der Riet, W., (1978) , *Green Mango Aachar*, National Food Research Institute, Council for Scientific and Industrial Research, Pretoria, South Africa.

[108]Fellows, P., (1997), *Traditional Foods*, Intermediate Technology Publications, UK

[109] Lal, G., Siddappa, G.S. and Tandon, G.L. ,(1986*), Preservation of Fruits and Vegetables*, Indian Council of Agricultural Research, India

[110] Campbell-Platt, G. (1987). *Fermented Foods of the World - a dictionary and guide.* Butterworths, London, UK

[111]Karki,T., (1986), *Some Nepalese Fermented Foods and Beverages*, in "Traditional Foods: Some Products and Technologies", Central Food Technological Research Insitute, India

[112]Aidoo, K.E., (1986), *Lesser-known fermented plant foods*, Tropical Science, 1986, 26, 249-258

[113]Jones, A., (1994), *The Ancient Art of Biotechnology*, Food Chain, 13, Intermediate Technology, UK

[114]Karki,T., (1986), *Some Nepalese Fermented Foods and Beverages*, in "Traditional Foods: Some Products and Technologies", Central Food Technological Research Insitute, India

[115]Karki, T., (1986), *Some Nepalese Fermented foods and beverages*' in "Traditional Foods", Central Food Technological Research Institute, India

[116]Aidoo, K. E., (1992) in Applications of Biotechnology to Traditional Fermented Foods, National Academy Press, Washington, USA.

[117] Steinkraus K H, (1996). *Handbook of Indigenous Fermented Foods*. Marcel Decker Inc, New York.

[118] Steinkraus K H, (1996). *Handbook of Indigenous Fermented Foods*. Marcel Decker Inc, New York.

[119]Karki,T., (1986), *Some Nepalese Fermented Foods and Beverages*, in "Traditional Foods: Some Products and Technologies", Central Food Technological Research Insitute, India

[120] Steinkraus K H, (1996). *Handbook of Indigenous Fermented Foods*. Marcel Decker Inc, New York.

[121]Makayama, T., (1957), Nippon Nogeikagakukaishi, 23, 497, Japan

[122]Berry, (1998), *Kanji,* Food Chain, 23, Intermediate Technology, UK

[123]Shah, F.H., (1986), *Fermented Foods of Pakistan,* in Traditional Foods: Some Products and Technologies, Central Food Technological Research Institute, India.

[124]Dirar, H. (1992) in *Applications of Biotechnology to Traditional Fermented Food*, National Academy Press, Washington, USA

[125]Harper, D.B. and Collins, M.A., (1992), *Leaf and Seed Fermentations of Western Sudan*, in "Applications of Biotechnology to Traditional Fermented Foods", National Academy Press, USA

[126]Aidoo, K.E., (1986), *Lesser-known fermented plant foods*, Tropical Science, 1986, 26, 249-258

[127]Jones, A., Hidellage, V., Wedgewood, H., Appleton, H. And Battcock, M, (1996), *Food Processing,* in "Biotechnology, Building on Farmers' Knowledge", Edited by Bunders, J., Haverkort, B., and Hiemstra, W., Macmillan, UK

[128]Menea, K.B. and Bishosha, A.K., (1992) *Transformation of Cassava Leaves by Lokele people in the Republic of Zaire.* ETC correspondence

[129] Steinkraus, K. H. (1996). *Handbook of Indigenous Fermented Foods.* Marcel Decker Inc, New York.

[130] Anon, *Vinegar: Sour Sap that Spells Sweet Gains*, Industrial Fermentation Program, National Institute of Science and Technology, Phillipines.

[131] Fellows, P., (1997), *Traditional Foods*, Intermediate Technology Publications, UK

[132] Jayawardena, J., (1977), *Production of vinegar*, Industrial Development Board, Sri Lanka

[133] Paivoke, A., Adams, M.R., and Twiddy, D.R., (1984), *Nipa Palm Vinegar in Papua New Guinea,* Process Biochemistry, Vol 19, no3, UK

[134] Kordylas, J.M., (1990), *Processing and Preservation of Tropical and Subtropical Foods,* Macmillan, UK

[135] Azami, S., (1994), *Pickles of Bangladesh*, Intermediate Technology, UK

[136] Wood G.A.R., (1975), *Cocoa,* Longman, Tropical Agriculture Series

[137] Beckett, S.T., (1988), *Industrial Chocolate: Manufacture and Use,* Blackie, London, UK

[138]Carr, J.G., (1985), *Tea, Coffee and Cocoa*, in "Microbiology of Fermented Foods", Wood, B.J.B., Elsevier, UK

[139] Minifie, B.W., (1980), *Chocolate, Cocoa and Confectionery* Science and Technology, AVI, USA

[140] Wood G.A.R., (1975), *Cocoa,* Longman, Tropical Agriculture Series

[141] Cakebread, S., Sugar and Chocolate Confectionary, UK

[142] Wood G.A.R., (1975), *Cocoa,* Longman, Tropical Agriculture Series

[143] Glossop, E., (1993), *Cocoa Processing,* Tropical Development Research Institute, UK

[144] Gates, J., (1990), *Growing Role of Grinding,* Coffee and Cocoa International

[145] Beckett, S.T., (1988), *Industrial Chocolate: Manufacture and Use,* Blackie, London, UK

[146] Ganesan, V., & George, T.P., (1990), Processing of Cocoa Beans at Small Scale Level

[147] Wood G.A.R., (1975), *Cocoa,* Longman, Tropical Agriculture Series

[148]Wellman, F.L., (1961), *Coffee*, Leonard Hill Books, UK

[149]Pedersen, C.S. and Breed, R.B., (1946), *Fermentation of Coffee*, Food Research 11, pp99-106

[150]Vaughn R.H., De Camargo L., Falanghe, H., Mello-Ayres, G. And Serzedello, A. (1958), *Observations on the Microbiology of the Coffee Fermentation in Brazil*, Food Technology, 12 Abstract 173

[151]Hilmer, A.F., Lum, N.A., and Dela Cruz, A.S., (1965), *Bacteria responsible for mucilage-layer decompositionin Kona coffee cherries,* Applied Microbiology, 13, pp201-7

[152]Agate, A.D. and Bhat, J.V., (1966), *Role of pectinolytic yeasts in the depradation of the mucilage layer*, Applied Microbiology, 14, pp256-60

[153] Anon, (1995), Coffee Technical Brief, Intermediate Technology.

[154] Fellows, P., (1997), *Traditional Foods*, Intermediate Technology Publications, UK

[155]Carr, J.G., (1985), *Tea, Coffee and Cocoa*, in "Microbiology of Fermented Foods", Wood, B.J.B., Elsevier, UK

[156] Fellows, P., (1997), *Traditional Foods*, Intermediate Technology Publications, UK

[157] Cowan, C. (1998), *Irelands Traditional Foods*. National food Centre, Ireland. (To be published in 1998).

[158] Broerse, J.E.W. and Visser, B., (1996), Assessing the Potential, in Biotechnology: Building on Farmers' Knowledge, edited by Bunders, J., Haverkort, B. and Hiemstra, W., (1996), Macmillan Education Ltd, UK).

[159] Broerse, J.E.W. and Visser, B., (1996), Assessing the Potential, in Biotechnology: Building on Farmers' Knowledge, edited by Bunders, J., Haverkort, B. and Hiemstra, W., (1996), Macmillan Education Ltd, UK).

[160] Okafor, N. (1992), Commercialisation of fermented Foods in Sub-Saharan Africa, in Application

[161] Nout, M.J.R., (1992) in Applications of Biotechnology to Traditional Fermented Foods, report of an Ad Hoc Panel of the Board on Science and Technology for International Development, (1992), National Academy Press, Washington D.C., USA

[162] Fellows, P., (1997), *Traditional Foods*, Intermediate Technology Publications, UK